Lectures on Differential Geometry with Maple

This text is designed to update the Differential Geometry course by making it more relevant to today's students. This new approach emphasizes applications and computer programs aimed at twenty-first century audiences. It is intended for mathematics students, applied scientists, and engineers who attempt to integrate differential geometry techniques in their work or research.

The course can require students to carry out a daunting amount of time-consuming hand computations like the computation of the Christoffel symbols. As a result, the scope of the applied topics and examples possible to cover might be limited. In addition, most books on this topic have only a scant number of applications.

This book is meant to evolve the course by including topics that are relevant to students. To achieve this goal, this book uses numerical, symbolic computations, and graphical tools as an integral part of the topics presented. This book provides students with a set of Maple/MATLAB® programs that will enable them to explore the course topics visually and in depth. These programs facilitate topic and application integration and provide the student with visual reinforcement of the concepts, examples, and exercises of varying complexity.

This unique text will empower students and users to explore in depth and visualize the topics covered, while these programs can be easily modified for other applications or other packages of numerical/symbolic languages. The programs are available to download for instructors and students using this book for coursework.

Mayer Humi is a Professor of Mathematics at Worcester Polytechnic University. He holds a Ph.D. in Mathematical Physics and Mathematical Modeling. He is an Associate Editor of the *International Journal of Differential Equations* and has published over 90 journal papers. His research focuses on the development and application of mathematical methods to atmospheric research and satellite orbits. Other research topics include mathematical physics, celestial mechanics, atmospheric flow, Lie groups, and differential equations.

Textbooks in Mathematics

Series editors:

Al Boggess, Kenneth H. Rosen

Measure Theory and Fine Properties of Functions, Second Edition
Lawrence Craig Evans and Ronald F. Gariepy

Set Theory: An Introduction to Axiomatic Reasoning
Robert André

Introduction to Differential and Difference Equations Through Modeling
William P. Fox and Robert E. Burks

Abstract Algebra, Third Edition An Interactive Approach
William Paulsen

Elements of Algebraic Topology, Second Edition
James R. Munkres, Steven G. Krantz, and Harold R. Parks

One Complex Variable from the Several Variable Point of View
Peter V. Dovbush and Steven G. Krantz

Math Anxiety How to Beat It
Brian Cafarella

Lectures on Differential Geometry with Maple
Mayer Humi

For more information about this series, please visit:
https://www.routledge.com/Textbooks-in-Mathematics/book-series/
CANDHTEXBOOMTH

Lectures on Differential Geometry with Maple

Mayer Humi

CRC Press

Taylor & Francis Group

Boca Raton London New York

CRC Press is an imprint of the
Taylor & Francis Group, an **informa** business

A CHAPMAN & HALL BOOK

Designed cover image: Mayer Humi

MATLAB® and Simulink® are trademarks of The MathWorks, Inc. and are used with permission. The MathWorks does not warrant the accuracy of the text or exercises in this book. This book's use or discussion of MATLAB® or Simulink® software or related products does not constitute endorsement or sponsorship by The MathWorks of a particular pedagogical approach or particular use of the MATLAB® and Simulink® software.

First edition published 2026
by CRC Press
2385 Executive Center Drive, Suite 320, Boca Raton, FL 33431
and by CRC Press
4 Park Square, Milton Park, Abingdon, Oxon, OX14 4RN

CRC Press is an imprint of Taylor & Francis Group, LLC
© 2026 Mayer Humi

Reasonable efforts have been made to publish reliable data and information, but the author and publisher cannot assume responsibility for the validity of all materials or the consequences of their use. The authors and publishers have attempted to trace the copyright holders of all material reproduced in this publication and apologize to copyright holders if permission to publish in this form has not been obtained. If any copyright material has not been acknowledged please write and let us know so we may rectify in any future reprint.

ISBN: 978-1-032-95978-8 (hbk)
ISBN: 978-1-032-95750-0 (pbk)
ISBN: 978-1-003-58742-2 (ebk)

DOI: 10.1201/9781003587422

Typeset in CMR10
by codeMantra

Contents

Preface

There is a long list of excellent books on classical and modern differential geometry. However, in many cases, these books treat differential geometry as a (pure) mathematical discipline without regard to its important applications in science and engineering or provide only a scant number of "trivial" applications.

I taught a course on differential geometry for many years at WPI. In my experience, this course required students to carry out a daunting amount (and time consuming) hand computations which are prone to mistakes even for some of the most simple topics in the course. In particular, the computation of the Christoffel Symbols was a "road block" for many students.

This situation limited severely the interest of students in this course, and it was impossible to convey to the audience the rich scope of the topics covered and their current applications. As a result, the number of the applied topics and examples that it was possible to cover was limited.

The motivation and objective of this book is to remedy these problems, update this course, and include topics that are relevant for a twenty-first century audience. It is intended for students, applied scientists, and engineers who attempt to integrate differential geometry techniques in their work or research. To this end, this book attempts to cut through the theoretical and computational bottlenecks that one must overcome for practical applications. To achieve this goal, this book uses numerical and symbolic computations as an integral part of the topics presented and provides a set of Maple/MATLAB® programs to facilitate this integration. This integration will empower students and users to explore visually and in depth the topics covered with a minimum amount of work. These programs can be easily modified for other applications or other packages of numerical/symbolic languages. The programs are available to download from the following url: http://users.wpi.edu/mhumi/DGMaple.tar.gz

1

Geometry of Curves in 3D

1.1 Curves and Parameterizations

In parametric form, a curve in three dimensions is a mapping

$$\mathbf{x} : U \subset R \to R^3,$$

where U is a subset of R (real numbers) and R^3 is the three-dimensional Euclidean Space.

Example 1.1: A straight line $R \to R^3$ is given by

$$\mathbf{x}(t) = (a_1 + m_1 t, a_2 + m_2 t, a_3 + m_3 t),$$

where the a_i, m_i, $i = 1, 2, 3$ are constants.

Example 1.2: An ellipse in the x-y plane is parameterized as

$$\mathbf{x}(t) = (a \cos \theta, b \sin \theta, 0).$$

To see this, we note that $\frac{x}{a} = \cos \theta$, $\frac{y}{b} = \sin \theta$, and therefore, this curve satisfies

$$\frac{x^2}{a^2} + \frac{y^2}{b^2} = 1.$$

Similarly, a hyperbola $\frac{x^2}{a^2} - \frac{y^2}{b^2} = 1$ can be represented by

$$\mathbf{x}(t) = (a \cosh \theta, b \sinh \theta, 0).$$

However, we should note that other parameterizations are possible, e.g for the ellipse

$$\mathbf{x} = \left(x, \pm b \sqrt{1 - \frac{x^2}{a^2}}, 0 \right).$$

In this formula, x is the parameter. Similar parameterizations (in terms of x) can be derived for the hyperbola.

Example 1.3: A circular helix with constant torsion (see Figure 1.1) is given by

$$\mathbf{x}(t) = (a \cos t, a \sin t, ct).$$

DOI: 10.1201/9781003587422-1

1

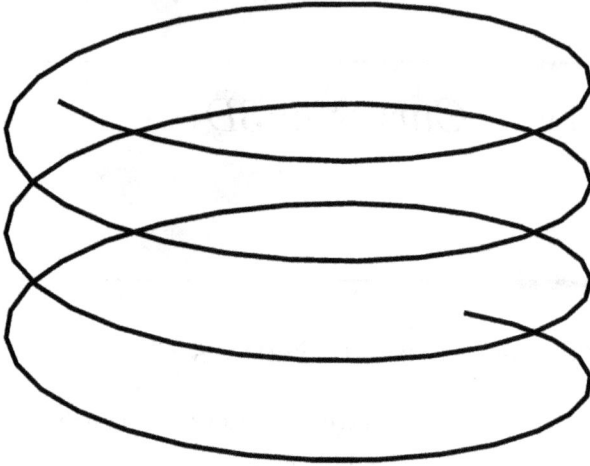

FIGURE 1.1
A graph of the circular Helix.

Example 1.4: The Folium of Descartes is a plane curve (see Figure 1.2). The Cartesian representation of this curve is

$$x^3 + y^3 = 3xy. \tag{1.1}$$

To drive a parametric representation of this curve, we let $t = y/x$, substitute for y in (1.1) and solve for x to obtain that

$$x = \frac{3t}{1 + t^3}.$$

Hence, the parametric representation of this curve in terms of t is

$$\mathbf{x}(t) = \left(\frac{3t}{1 + t^3}, \frac{3t^2}{1 + t^3}, 0 \right). \tag{1.2}$$

This curve does not have a unique tangent at the origin.

Remark 1.1: A more general form for Folium of Descartes is

$$x^3 + y^3 = 3\alpha xy,$$

where α is a constant.

We observe that (usually) there are infinitely many ways to parameterize a curve, e.g., other possible parameterizations of the ellipse are

$$\mathbf{x}(t) = (a \cos n\theta, b \sin n\theta, 0), \quad n = 1, 2, \ldots,$$

with proper restrictions on the range of θ.

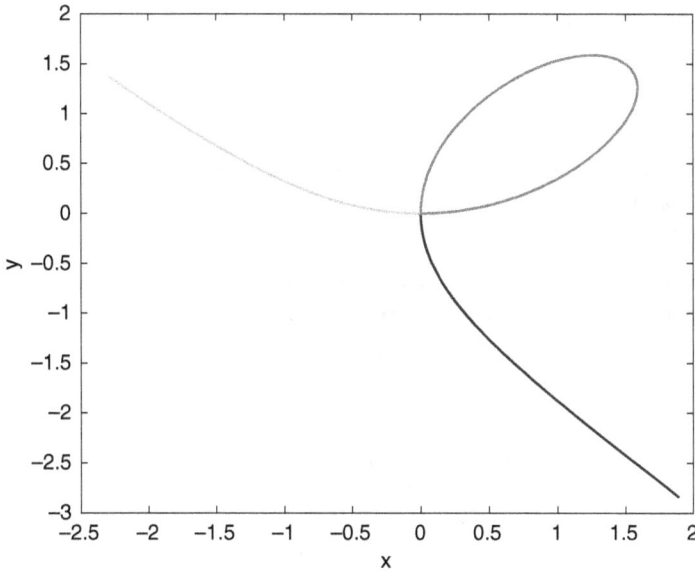

FIGURE 1.2
A graph of the Folium of Descartes.

However, out of all these parameterizations, there is one that is **intrinsic** (or generic) to the curve itself. This parameterization consists of choosing an arbitrary fixed point on the curve and then using the distance s along the curve from this point to parameterize the points on the curve. While this parameterization might not be practical in many cases, it is most natural from a theoretical point of view. Formulas using s can be converted "easily" to other parameterizations using the chain rule.

Review 1.1: The infinitesimal distance along a plane curve is

$$ds^2 = dx^2 + dy^2.$$

Therefore if $x = x(t)$ and $y = y(t)$, then

$$\frac{ds}{dt} = \sqrt{\left(\frac{dx}{dt}\right)^2 + \left(\frac{dy}{dt}\right)^2}.$$

Hence

$$s(t) = \int_{t_0}^{t_1} \sqrt{\left(\frac{dx}{dt}\right)^2 + \left(\frac{dy}{dt}\right)^2} \, dt.$$

In particular, if we use x as a parameter i.e $y = y(x)$, then

$$s(x) = \int_{x_0}^{x_1} \sqrt{1 + \left(\frac{dy}{dx}\right)^2} \, dx.$$

Note also that along the curve

$$\left(\frac{dx}{ds}\right)^2 + \left(\frac{dy}{ds}\right)^2 = 1.$$

The presence of the square root in the integrals above render them (in general) difficult to compute analytically. However, from a practical point of view, we shall need only $\frac{ds}{dt}$, etc. which are readily available.

The situation in three dimensions is similar to the two-dimensional case in particular from Pythagoras theorem in R^3 we have,

$$ds^2 = (dx)^2 + (dy)^2 + (dz)^2.$$

Hence

$$\left(\frac{dx}{ds}\right)^2 + \left(\frac{dy}{ds}\right)^2 + \left(\frac{dz}{ds}\right)^2 = 1.$$

Example 1.5: Compute the arc-length of the circular Helix (see Figure 1.1).

$$\mathbf{x} = (a \cos t, a \sin t, ct).$$

Solution:

$$\frac{dx}{dt} = -a \sin t, \quad \frac{dy}{dt} = a \cos t, \quad \frac{dz}{dt} = c.$$

Therefore

$$\left(\frac{ds}{dt}\right)^2 = a^2 + c^2 = b^2.$$

Hence

$$s = bt + \text{constant}.$$

Example 1.6: Compute the arc-length of the Catenary. The Catenary is a curve that represents the shape of a hanging chain under gravity (see Figure 1.3). We derive its equation in the Appendix.

$$\mathbf{x} = (t, a \cosh \frac{t}{a}, 0).$$

Solution:

$$\frac{dx}{dt} = 1, \quad \frac{dy}{dt} = \sinh \frac{t}{a}.$$

Therefore

$$\left(\frac{ds}{dt}\right)^2 = 1 + \sinh^2 \frac{t}{a} = \cosh^2 \frac{t}{a}.$$

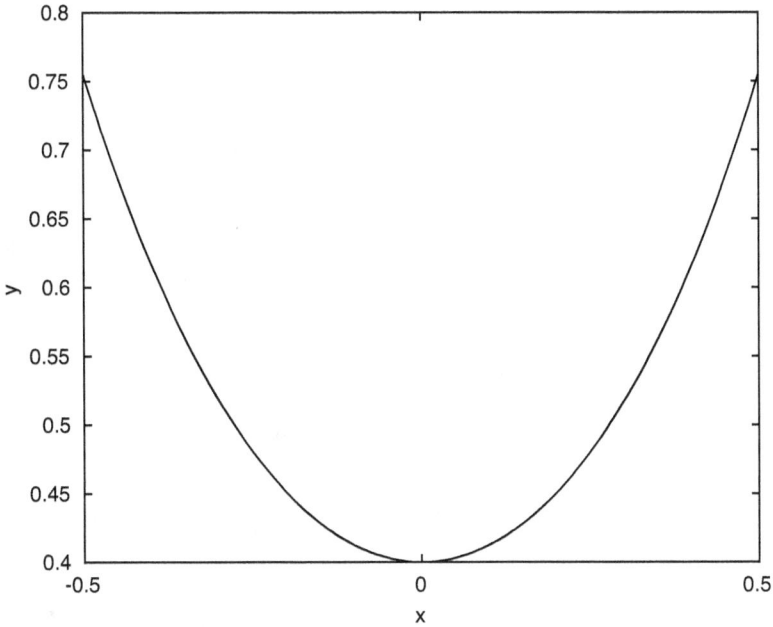

FIGURE 1.3
A picture of a Catenary.

Hence, using the "origin" (viz $t = 0$ which corresponds to $(0,1,0)$ on the curve) as the "origin point" to measure distances from

$$s = \int \cosh \frac{t}{a} dt = a \sinh \frac{t}{a}.$$

Remark 1.2: For a curve in R^2 whose parametric representation in polar coordinates is $(r(t), \theta(t))$ we have $x = r(t)\cos(\theta(t))$ and $y(t) = r(t)\sin(\theta(t))$. Therefore

$$\left(\frac{ds}{dt}\right)^2 = \left(\frac{dx}{dt}\right)^2 + \left(\frac{dy}{dt}\right)^2 = \left(\frac{dr}{dt}\right)^2 + r^2\left(\frac{d\theta}{dt}\right)^2.$$

Example 1.7: Calculate the arc length of the Archimedean spiral whose representation in polar coordinates (r, θ) is $(a\theta, \theta)$ (viz. θ is the parameter and $r(\theta) = a\theta$).

Solution: Applying the formula derived above to this curve, we have

$$s = \int_{\theta_0}^{\theta} \sqrt{(\frac{dr}{d\theta})^2 + r^2} \, d\theta.$$

In this case $\frac{dr}{d\theta} = a$, therefore

$$s = \int_{\theta_0}^{\theta} \sqrt{a^2 + a^2\theta^2}\, d\theta = a \int_{\theta_0}^{\theta} \sqrt{1 + \theta^2}\, d\theta.$$

To compute the last integral, we use the integration by parts formula

$$\int u'v\,dx = uv - \int uv'\,dx,$$

with $u' = 1$, $v = \sqrt{1 + \theta^2}$ which imply that $u = \theta$ and $v' = \theta(1 + \theta^2)^{-1/2}$.
Hence

$$
\begin{aligned}
\int \sqrt{1 + \theta^2}\, d\theta &= \theta\sqrt{1 + \theta^2} - \int \frac{\theta^2}{\sqrt{1 + \theta^2}}\, d\theta && \text{(1.3)} \\
&= \theta\sqrt{1 + \theta^2} - \int \frac{1 + \theta^2 - 1}{\sqrt{1 + \theta^2}}\, d\theta \\
&= \theta\sqrt{1 + \theta^2} - \int \sqrt{1 + \theta^2}\, d\theta + \int \frac{d\theta}{\sqrt{1 + \theta^2}}.
\end{aligned}
$$

Therefore

$$2 \int \sqrt{1 + \theta^2}\, d\theta = \theta\sqrt{1 + \theta^2} + \int \frac{d\theta}{\sqrt{1 + \theta^2}}.$$

To compute the integral on the right-hand side of the last equation, we make
a change of variables.

$$\theta = \sinh t, \quad 1 + \theta^2 = 1 + \sinh^2 t = \cosh^2 t, \quad d\theta = \cosh t\, dt.$$

We obtain

$$\int \frac{d\theta}{\sqrt{1 + \theta^2}} = \int \frac{\cosh t}{\cosh t}\, dt = t = \operatorname{arcsinh}(\theta).$$

Hence

$$2 \int \sqrt{1 + \theta^2}\, d\theta = \theta\sqrt{1 + \theta^2} + \operatorname{arcsinh}(\theta).$$

Finally

$$s = \frac{1}{2}[\theta\sqrt{1 + \theta^2} + \operatorname{arcsinh}(\theta)],$$

(where we assumed $\theta_0 = 0$).

1.2   Properties of Curves in R^3

How one characterize the local properties of a curve? In other words, if we are
given two curves passing through the same point in space what makes them
different.

1.2.1   The Local Frame

We want to attach to each point on the curve an orthogonal frame of vectors.

Geometrically, the first item that comes to mind is the tangent vector to the curve

$$\mathbf{T}(s) = \frac{d\mathbf{x}}{ds}.$$

We observe that by using the parameter s, the length of $\mathbf{T}$ at each point of the curve is 1. In fact,

$$|\mathbf{T}|^2 = \mathbf{T} \cdot \mathbf{T} = (\frac{dx}{ds}, \frac{dy}{ds}, \frac{dz}{ds}) \cdot (\frac{dx}{ds}, \frac{dy}{ds}, \frac{dz}{ds}) = \left(\frac{dx}{ds}\right)^2 + \left(\frac{dy}{ds}\right)^2 + \left(\frac{dz}{ds}\right)^2 = 1.$$

Observe that this holds only when we parameterize the curve in terms of s. For other parameterizations, we have by the chain rule

$$\bar{\mathbf{T}}(t) = \frac{d\mathbf{x}}{dt} = \frac{d\mathbf{x}}{ds}\frac{ds}{dt} = \mathbf{T}(s)\frac{ds}{dt}.$$

Thus, the unit tangent vector at t can be computed using the following formula,

$$\mathbf{T}(t) = \frac{\frac{d\mathbf{x}}{dt}}{\frac{ds}{dt}}. \tag{1.4}$$

Since $\mathbf{T}(s) \cdot \mathbf{T}(s) = 1$ we obtain by differentiation

$$\mathbf{T}(s) \cdot \frac{d\mathbf{T}(s)}{ds} = 0.$$

Hence, the vector $\bar{\mathbf{p}}(s) = \frac{d\mathbf{T}(s)}{ds}$ is orthogonal to $\mathbf{T}(s)$ (unless $\mathbf{T}(s)$ is the zero vector). The unit vector in the direction of $\bar{\mathbf{p}}$ i.e.

$$\mathbf{p}(s) = \frac{\bar{\mathbf{p}}(s)}{|\bar{\mathbf{p}}(s)|} = \frac{\frac{d\mathbf{T}(s)}{ds}}{\left|\frac{d\mathbf{T}(s)}{ds}\right|},$$

is called the **Principal Normal** to the curve C at the point P and

$$\kappa = \left|\frac{d\mathbf{T}(s)}{ds}\right|,$$

is referred to as the **curvature of the curve at s.** Hence

$$\frac{d\mathbf{T}(s)}{ds} = \frac{d^2\mathbf{x}(s)}{ds^2} = \kappa\mathbf{p}(s).$$

Thus,

$$\kappa(s) = \sqrt{\frac{d\mathbf{T}(s)}{ds} \cdot \frac{d\mathbf{T}(s)}{ds}} = \sqrt{\frac{d^2\mathbf{x}(s)}{ds^2} \cdot \frac{d^2\mathbf{x}(s)}{ds^2}}.$$

Finally, we introduce the **binormal vector** using the vector product

$$\mathbf{b} = \mathbf{T} \times \mathbf{p}.$$

The three unit vectors $\mathbf{T}, \mathbf{p}, \mathbf{b}$ form an orthogonal frame at each point of a curve C.

To see how the vectors in the frame $(\mathbf{T},\mathbf{p},\mathbf{b})$ change along the curve, we shall express their derivatives with respect to s using the vectors of this frame.

In the following, dots represent differentiation with respect to s.

Since $|\dot{\mathbf{T}}(s)| = \kappa(s)$ we have from the definition of $\mathbf{p}$ that

$$\dot{\mathbf{T}} = \kappa\mathbf{p}.$$

Since $\mathbf{b}$ is of unit length and is orthogonal to $\mathbf{T}$ we have

$$\mathbf{b} \cdot \mathbf{b} = 1, \quad \mathbf{b} \cdot \mathbf{T} = 0.$$

Differentiating the first relation yields

$$\mathbf{b} \cdot \dot{\mathbf{b}} = 0,$$

i.e $\dot{\mathbf{b}}$ is orthogonal to $\mathbf{b}$. Differentiating the second relation we obtain

$$\dot{\mathbf{b}} \cdot \mathbf{T} = -\mathbf{b} \cdot \dot{\mathbf{T}} = -\kappa\mathbf{b} \cdot \mathbf{p} = 0.$$

Hence $\dot{\mathbf{b}}$ is orthogonal to $\mathbf{T}$ and $\mathbf{b}$; therefore, it must be a vector in the direction of $\mathbf{p}$, i.e

$$\dot{\mathbf{b}} = -\tau\mathbf{p},$$

where $\tau(s)$ is referred to as the curve **torsion** at s. (The geometrical justification for this name will be explained in the next subsection.)

Finally, since $\mathbf{p}$ is of unit length it follows that $\dot{\mathbf{p}}$ is orthogonal to $\mathbf{p}$, and therefore, it can be expressed as a linear combination of the vectors $\mathbf{T}$, $\mathbf{b}$ viz.

$$\dot{\mathbf{p}} = \alpha\mathbf{T} + \beta\mathbf{b}.$$

To compute α and β, we invoke the orthogonality of $\mathbf{p}$ and $\mathbf{T}$, $\mathbf{b}$

Since $\mathbf{p} \cdot \mathbf{T} = 0$ it follows that

$$\dot{\mathbf{p}} \cdot \mathbf{T} + \mathbf{p} \cdot \dot{\mathbf{T}} = 0.$$

Hence

$$\alpha = \dot{\mathbf{p}} \cdot \mathbf{T} = -\mathbf{p} \cdot \dot{\mathbf{T}} = -\kappa.$$

Similarly,

$$\beta = \dot{\mathbf{p}} \cdot \mathbf{b} = -\mathbf{p} \cdot \dot{\mathbf{b}} = \tau.$$

Combining all these results, we obtain the following system of differential equations

$$
\begin{pmatrix} \dot{\mathbf{T}} \\ \dot{\mathbf{p}} \\ \dot{\mathbf{b}} \end{pmatrix} = \begin{pmatrix} 0 & \kappa & 0 \\ -\kappa & 0 & \tau \\ 0 & -\tau & 0 \end{pmatrix} \begin{pmatrix} \mathbf{T} \\ \mathbf{p} \\ \mathbf{b} \end{pmatrix}.
\tag{1.5}
$$

This system of equations is referred to as **Frenet formulas** in R^3. In general, one can construct numerically a curve in three dimensions with given $\kappa(s)$ and $\tau(s)$ by solving the system of equation (1.5) and then integrate $\mathbf{T}(s)$ to find $\mathbf{x}(s)$.

There is a generalization of these formulas to curves in R^n

Theorem 1.1: Let C be a curve in R^n. There exists along C a set of orthogonal unit vectors $\mathbf{E} = (\mathbf{e_1}(s), \ldots, \mathbf{e_n}(s))^T$ which satisfy

$$
\frac{d\mathbf{E}}{ds} = F\mathbf{E},
$$

where F is a tridiagonal skew-symmetric matrix. The entries in F are referred to as "Frenet curvatures".

Theorem 1.2: Fundamental Theorem about Curves in R^n
Let $(k_1, \ldots, k_{n-1})$ be C^∞ functions $R \to R^n$ then for any point $p_0 \in R^n$ and a frame $\mathbf{E}(p_0)$ there is a unique curve passing through p_0 whose Frenet frame at this point is $\mathbf{E}(p_0)$ and its Frenet curvatures at this point are $(k_1(p_0), \ldots, k_{n-1}(p_0))$.

1.3 Curvature and Torsion

Frenet formulas characterize the local properties of a curve C in terms of two functions $\kappa(s)$ and $\tau(s)$. Although these two functions were introduced formally in the previous section, they characterize the local geometric properties of the curve.

Definition 1.1: The functions $\kappa(s)$ and $\tau(s)$ are called the **curvature** and **torsion** of the curve C respectively.

To justify this nomenclature, we consider several examples.

Example 1.8: Consider the circle

$$
\mathbf{x} = (r \cos \theta, r \sin \theta, 0).
$$

For this curve

$$ds^2 = [(-r\sin\theta)^2 + (-r\cos\theta)^2](d\theta)^2 = r^2(d\theta)^2.$$

Hence $s = r\theta$. Therefore

$$\mathbf{x} = (r\cos\frac{s}{r}, r\sin\frac{s}{r}, 0).$$

The tangent vector is

$$\mathbf{T} = \frac{d\mathbf{x}}{ds} = (-\sin\frac{s}{r}, \cos\frac{s}{r}, 0),$$

and

$$\dot{\mathbf{T}} = \frac{1}{r}(-\cos\frac{s}{r}, -\sin\frac{s}{r}, 0).$$

Therefore $\mathbf{p} = (-\cos\frac{s}{r}, -\sin\frac{s}{r}, 0)$ and $\kappa = 1/r$. Finally,

$$\mathbf{b} = \mathbf{T} \times \mathbf{p} = (0, 0, 1),$$

i.e $\mathbf{b}$ is a constant vector, and therefore $\dot{\mathbf{b}} = 0$ i.e $\tau = 0$. Thus, this plane curve has no torsion, and the curvature is $1/r$. As $r \to \infty$ the curvature goes to zero which is intuitively correct. (A straight line has zero curvature).

Definition 1.2: $\frac{1}{\kappa(\mathbf{P})}$ is called the radius of curvature of the curve at $\mathbf{P}$.

Theorem 1.3: A curve C with non-vanishing curvature is a plane curve if and only if $\tau(s) = 0$ identically.

Proof: If C is a plane curve, we can choose this plane to be the x-y plane. Since both $\mathbf{T}(s)$ and $\mathbf{p} = \frac{1}{\kappa}\frac{d\mathbf{T}}{ds}$ are in the x-y plane, it follows that $\mathbf{b}$ which is orthogonal to these two vectors has a constant direction (and length). Therefore, $\dot{\mathbf{b}} = 0$ and hence $\tau(s) = 0$ for all the point of C.

Conversely if $\tau = 0$ then $\mathbf{b}$ is a constant vector ($\dot{\mathbf{b}} = \mathbf{0}$). Now consider the function

$$F(s) = (\mathbf{x}(s) - \mathbf{x}(0)) \cdot \mathbf{b}.$$

Then

$$\frac{dF}{ds} = \frac{d\mathbf{x}(s)}{ds} \cdot \mathbf{b} = \mathbf{T} \cdot \mathbf{b} = 0.$$

Therefore, $F(s) = constant$ but obviously $F(0) = 0$ hence $F(s)$ is identically zero and all the points of the curve are in the plane

$$(\mathbf{y} - \mathbf{x}(0)) \cdot \mathbf{b} = 0,$$

whose normal is $\mathbf{b}$.

Thus when $\tau = 0$ the curve is in a plane and when the torsion $\tau \neq 0$ it is not a plane curve.

Example 1.9: The circular Helix parameterized in terms of s is

$$\mathbf{x} = \left(a \cos \frac{s}{b}, a \sin \frac{s}{b}, \frac{cs}{b} \right),$$

where $b^2 = a^2 + c^2$. The unit tangent is

$$\mathbf{T}(s) = \frac{d\mathbf{x}}{ds} = \left(-\frac{a}{b} \sin \frac{s}{b}, \frac{a}{b} \cos \frac{s}{b}, \frac{c}{b} \right),$$

and

$$\frac{d\mathbf{T}}{ds} = \left(-\frac{a}{b^2} \cos \frac{s}{b}, -\frac{a}{b^2} \sin \frac{s}{b}, 0 \right).$$

Therefore,

$$\mathbf{p} = \left(-\cos \frac{s}{b}, -\sin \frac{s}{b}, 0 \right),$$

and

$$\kappa(s) = \frac{a}{b^2}.$$

Hence

$$\mathbf{b} = \mathbf{T} \times \mathbf{p} = \left(\frac{c}{b} \sin \frac{s}{b}, -\frac{c}{b} \cos \frac{s}{b}, \frac{a}{b} \right).$$

To compute τ we differentiate $\mathbf{b}$

$$\frac{d\mathbf{b}}{ds} = \left(\frac{c}{b^2} \cos \frac{s}{b}, \frac{c}{b^2} \sin \frac{s}{b}, 0 \right),$$

and therefore (since $\frac{d\mathbf{b}}{ds} = -\tau \mathbf{p}$)

$$\tau = \frac{c}{b^2} = \frac{c}{a^2 + c^2}.$$

We see that the torsion is zero if $c = 0$ as expected.

General formulas for the curvature and torsion of a curve in **any** parameterization are as follows:

$$\kappa = \frac{|\mathbf{x}' \times \mathbf{x}''|}{(|\mathbf{x}'|)^3}, \tag{1.6}$$

$$\tau = \frac{\det(\mathbf{x}', \mathbf{x}'', \mathbf{x}''')}{(\mathbf{x}' \times \mathbf{x}'') \cdot (\mathbf{x}' \times \mathbf{x}'')}, \tag{1.7}$$

where primes denote differentiation with respect to the parameter t.

Remark 1.3: The triple product of three vectors $\mathbf{a}, \mathbf{b}, \mathbf{c}$

$$(\mathbf{a} \times \mathbf{b}) \cdot \mathbf{c} = \det(\mathbf{a}, \mathbf{b}, \mathbf{c}).$$

Observe that this product equals zero if two of these vectors are multiple of each other.

Proof: To prove (1.6), we note that

$$\mathbf{T} = \frac{d\mathbf{x}}{ds}, \quad \frac{d\mathbf{T}}{ds} = \frac{d^2\mathbf{x}}{ds^2} = \kappa\mathbf{p}.$$

Therefore,

$$\left|\mathbf{T} \times \frac{d\mathbf{T}}{ds}\right| = \kappa|\mathbf{T} \times \mathbf{p}| = \kappa,$$

since $\mathbf{T}$ and $\mathbf{p}$ are of unit length and orthogonal to each other. Hence

$$\kappa = \left|\frac{d\mathbf{x}}{ds} \times \frac{d^2\mathbf{x}}{ds^2}\right|.$$

When the curve is parameterized by some parameter t, we have

$$\frac{d\mathbf{x}}{dt} = \frac{d\mathbf{x}}{ds}\frac{ds}{dt}, \quad \frac{d^2\mathbf{x}}{dt^2} = \frac{d^2\mathbf{x}}{ds^2}\left(\frac{ds}{dt}\right)^2 + \frac{d\mathbf{x}}{ds}\frac{d^2s}{dt^2}. \tag{1.8}$$

It follows then that

$$\frac{d\mathbf{x}}{dt} \times \frac{d^2\mathbf{x}}{dt^2} = \left(\frac{d\mathbf{x}}{ds} \times \frac{d^2\mathbf{x}}{ds^2}\right)\left(\frac{ds}{dt}\right)^3. \tag{1.9}$$

Hence

$$\kappa = \frac{\left|\frac{d\mathbf{x}}{dt} \times \frac{d^2\mathbf{x}}{dt^2}\right|}{\left(\frac{ds}{dt}\right)^3}.$$

But

$$\frac{d\mathbf{x}}{dt} = \frac{d\mathbf{x}}{ds}\frac{ds}{dt}.$$

However, since $|\frac{d\mathbf{x}}{ds}| = 1$ it follows that

$$\left|\frac{ds}{dt}\right| = \left|\frac{d\mathbf{x}}{dt}\right|,$$

which finally yields

$$\kappa = \frac{\left|\frac{d\mathbf{x}}{dt} \times \frac{d^2\mathbf{x}}{dt^2}\right|}{(|\frac{d\mathbf{x}}{dt}|)^3}. \tag{1.10}$$

To prove the formula for τ we note that by definition of the vector $\mathbf{b}$ and Frenet formulas

$$\tau = -\frac{d\mathbf{b}}{ds} \cdot \mathbf{p} = -\frac{d}{ds}(\mathbf{T} \times \mathbf{p}) \cdot \mathbf{p}.$$

However, since $(\mathbf{T} \times \mathbf{p}) \cdot \mathbf{p} = 0$,

$$-\frac{d}{ds}(\mathbf{T} \times \mathbf{p}) \cdot \mathbf{p} = (\mathbf{T} \times \mathbf{p}) \cdot \frac{d\mathbf{p}}{ds}.$$

Therefore,

$$\tau = (\mathbf{T} \times \mathbf{p}) \cdot \frac{d\mathbf{p}}{ds}.$$

But

$$\mathbf{T} = \frac{d\mathbf{x}}{ds}, \quad \mathbf{p} = \frac{1}{\kappa}\frac{d\mathbf{T}}{ds} = \frac{1}{\kappa}\frac{d^2\mathbf{x}}{ds^2},$$

and

$$\frac{d\mathbf{p}}{ds} = \frac{d(\frac{1}{\kappa})}{ds}\frac{d^2\mathbf{x}}{ds^2} + \frac{1}{\kappa}\frac{d^3\mathbf{x}}{ds^3}.$$

Combining these expressions, we have

$$\tau = \left(\frac{d\mathbf{x}}{ds} \times \frac{1}{\kappa}\frac{d^2\mathbf{x}}{ds^2}\right) \cdot \left(\frac{d(\frac{1}{\kappa})}{ds}\frac{d^2\mathbf{x}}{ds^2} + \frac{1}{\kappa}\frac{d^3\mathbf{x}}{ds^3}\right).$$

However, the triple product

$$\left(\frac{d\mathbf{x}}{ds} \times (\frac{1}{\kappa}\frac{d^2\mathbf{x}}{ds^2}\right) \cdot \left(\frac{d(\frac{1}{\kappa})}{ds}\frac{d^2\mathbf{x}}{ds^2}\right) = 0.$$

Two of the vectors in this triple product are the same. Therefore,

$$\tau = \frac{1}{\kappa^2}\left(\frac{d\mathbf{x}}{ds} \times \frac{d^2\mathbf{x}}{ds^2}\right) \cdot \frac{d^3\mathbf{x}}{ds^3} = \frac{\det(\frac{d\mathbf{x}}{ds}, \frac{d^2\mathbf{x}}{ds^2}, \frac{d^3\mathbf{x}}{ds^3})}{\kappa^2}.$$

Recalling (1.10) (with s replacing t, we finally obtain

$$\tau = \frac{\det(\frac{d\mathbf{x}}{ds}\frac{d^2\mathbf{x}}{ds^2}, \frac{d^3\mathbf{x}}{ds^3})}{|\left(\frac{d\mathbf{x}}{ds} \times \frac{d^2\mathbf{x}}{ds^2}\right)|^2}.$$

A long calculation using the chain rule shows that this formula holds when s is replaced by an arbitrary parameter t for the curve.

Corollary 1.1: If the curve is a two-dimensional curve in the x-y plane and $y = y(x)$ (or in vector form $\mathbf{x} = (x, y(x), 0)$) then

$$\kappa = \frac{|y''|}{(1 + (y')^2)^{3/2}}, \quad \tau = 0.$$

1.3.1 Planes along a Curve

Definition 1.3: let P be a point on a curve C and $\mathbf{T}(s)$ the tangent to C at P. The plane normal to $\mathbf{T}(s)$ is called the **normal plane** to C at P.

Review 1.2: The equation of the plane at a point $\mathbf{x_0} = (x_0, y_0, z_0)$ with normal $\mathbf{N}$ is

$$(\mathbf{x} - \mathbf{x_0}) \cdot \mathbf{N} = 0.$$

Similarly, the planes normal to **p**, and **b** are called the rectifying and osculating planes, respectively, i.e.,

$$(\mathbf{y} - \mathbf{x}(P)) \cdot \mathbf{p}(P) = 0 \text{ is the Rectifying plane,}$$

and

$$(\mathbf{y} - \mathbf{x}(P)) \cdot \mathbf{b}(P) = 0 \text{ is the Osculating plane.}$$

Example 1.10: The equation of a circular helix is

$$\mathbf{x}(t) = (a\cos t, a\sin t, ct), \tag{1.11}$$

$$\bar{\mathbf{T}}(t) = \frac{d\mathbf{x}}{dt} = (-a\sin t, a\cos t, c). \tag{1.12}$$

But

$$\left(\frac{ds}{dt}\right)^2 = \left(\frac{dx}{dt}\right)^2 + \left(\frac{dy}{dt}\right)^2 + \left(\frac{dz}{dt}\right)^2 = a^2 + c^2 = b^2.$$

Therefore using (1.4) the unit tangent vector at t is

$$\mathbf{T}(t) = (-\frac{a}{b}\sin t, \frac{a}{b}\cos t, \frac{c}{b}).$$

The normal plane to the helix at the point $(a,0,0)$ (i.e. $t = 0$) is

$$(\mathbf{x} - (a,0,0)) \cdot (0, a, c) = 0,$$

i.e

$$ay + cz = 0.$$

1.3.2   Contact

Definition 1.4: We say that two curves $\mathbf{x}(t)$ and $\mathbf{y}(r)$ have a contact of order n at a point P if for appropriate values of s_0 and r_0

$$P = \mathbf{x}(t_0) = \mathbf{y}(r_0),$$

and

$$\frac{d^k\mathbf{x}}{dt^k}(t_0) = \frac{d^k\mathbf{y}}{dr^k}(r_0), \quad k = 1, \ldots n,$$

and $\frac{d^{n+1}\mathbf{x}}{dt^{n+1}}(t_0) \neq \frac{d^{n+1}\mathbf{y}}{dr^{n+1}}(r_0)$.

Example 1.11: The two circles $(x-2)^2 + y^2 = 4$ and $(x-1)^2 + y^2 = 1$ have a contact of order one at $(0,0)$ since they have the same tangent at this point.

1.4 Helices

Definition 1.5: A curve is called a helix if its tangent makes a constant angle with a fixed line in space.

For the circular helix (1.11), we showed that $\mathbf{T}$ is given by (1.12), and therefore, the cosine of the angle ϕ between the tangent and the z-axis is

$$\cos\phi = \mathbf{T}\cdot(0,0,1) = \frac{c}{b},$$

which is a constant. A general characterization of helices in terms of their curvature and torsion is given by the following theorem.

1.4.1 Lancret's Theorem

A curve with non-vanishing curvature is a helix if and only if $\frac{\tau(s)}{\kappa(s)}$ is a fixed constant (independent of the point on the curve).

Proof: A. Assume that the curve is a general helix. By definition there exists a unit vector $\mathbf{c}$ so that

$$\mathbf{c}\cdot\mathbf{T}(s) = \cos(\theta),$$

where $\mathbf{T}$ is the tangent to the curve and θ is a fixed angle. Differentiating this equation we have

$$\mathbf{c}\cdot\frac{d\mathbf{T}}{ds} = 0.$$

but $\frac{d\mathbf{T}}{ds} = \kappa\mathbf{p}$. Therefore since $\kappa \neq 0$ we have $\mathbf{c}\cdot\mathbf{p} = 0$. It follows then that $\mathbf{c}$ is in the plane generated by $\mathbf{T}$ and $\mathbf{b}$. Since $\mathbf{T}$ and $\mathbf{b}$ are orthogonal to each other we deduce that

$$\mathbf{c}\cdot\mathbf{b} = \sin(\theta).$$

Differentiating $\mathbf{c}\cdot\mathbf{p} = 0$ and using Frenet equation for $\frac{d\mathbf{p}}{ds}$, we obtain

$$\mathbf{c}\cdot\frac{d\mathbf{p}}{ds} = \mathbf{c}\cdot(-\kappa\mathbf{T}+\tau\mathbf{b}) = (-\kappa\cos(\theta)+\tau\sin(\theta)) = 0.$$

Hence, $\frac{\tau(s)}{\kappa(s)}$ is a constant .

B. Assume that $\frac{\tau(s)}{\kappa(s)}$ is a constant i.e. $\tau = \alpha\kappa$ and show that the curve is a helix. By Frenet formulas

$$\alpha\frac{d\mathbf{T}}{ds} + \frac{d\mathbf{b}}{ds} = (\alpha\kappa - \tau)\mathbf{p} = \mathbf{0}.$$

Hence(by integration)
$$\alpha \mathbf{T} + \mathbf{b} = \mathbf{c},$$

where $\mathbf{c}$ is a constant vector. It follows then that

$$\mathbf{c} \cdot \mathbf{T} = (\alpha \mathbf{T} + \mathbf{b}) \cdot \mathbf{T} = \alpha,$$

which proves our statement.

1.5   Plane Curves

From a geometric point of view, a plane curve is characterized by its arc length s and its curvature κ $(\tau = 0)$. We introduce therefore the following definition:

Definition 1.6: A plane curve that is defined by a relation between s and κ is said to be defined by the **natural equation of the curve**

To obtain additional geometrical interpretation for the curvature for these curves, we consider a plane curve $\mathbf{x}(s)$ and let θ be the angle between the x-axis and the tangent to the curve at $\mathbf{x}(s)$. (See Figure 1.4). We then have

$$\mathbf{T} = (\cos\theta, \sin\theta), \quad \frac{d\mathbf{T}}{d\theta} = (-\sin\theta, \cos\theta),$$

and hence $\mathbf{T} \cdot \frac{d\mathbf{T}}{d\theta} = 0$. This implies that $\mathbf{T}$ is orthogonal to $\frac{d\mathbf{T}}{d\theta}$. Furthermore, since the binormal to the curve is in the z-direction, we infer that

$$\frac{d\mathbf{T}}{d\theta} = \pm\mathbf{p}.$$

(due to the fact that $\frac{d\mathbf{T}}{d\theta}$ is of unit length orthogonal to $\mathbf{T}$).

Assuming that $\frac{d\mathbf{T}}{d\theta} = \mathbf{p}$ it follows that

$$\frac{d\mathbf{T}}{ds} = \frac{d\mathbf{T}}{d\theta}\frac{d\theta}{ds} = \kappa\mathbf{p},$$

viz.

$$\kappa = \frac{d\theta}{ds}.$$

Example 1.12: Consider a curve that is defined by

$$\frac{s}{\kappa} = c^2,$$

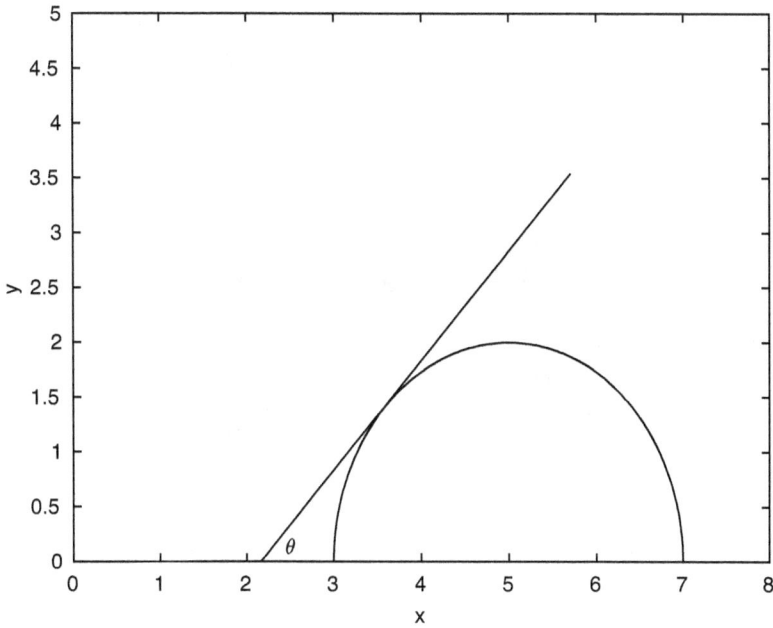

FIGURE 1.4
A two-dimensional curve and its tangent.

where c is a constant. Therefore, $\kappa = \frac{s}{c^2}$. Since $\frac{d\theta}{ds} = \kappa$ it follows that

$$\frac{d\theta}{ds} = \frac{s}{c^2}.$$

Integrating this equation with respect to s, we have

$$\theta = \int_0^s \frac{t}{c^2} dt = \frac{s^2}{2c^2}.$$

However since

$$\mathbf{T} = \left(\frac{dx}{ds}, \frac{dy}{ds} \right) = (\cos\theta, \sin\theta),$$

we have

$$dx = \cos\theta(s)ds, \quad dy = \sin\theta(s)ds.$$

Hence

$$x = \int_0^s \cos\theta(s)ds = \int_0^s \cos\left(\frac{s^2}{2c^2}\right)ds = \frac{c}{\sqrt{2}} \int_0^\phi \frac{\cos\nu}{\sqrt{\nu}} d\nu,$$

where $\nu = \frac{s^2}{2c^2}$. Similarly for y, we obtain

$$y = \int_0^\phi \frac{\sin\nu}{\sqrt{\nu}} d\nu.$$

These integrals for x, y are called "Frenet Integrals" and they appear in the treatment of refraction problems in optics. All the curves that are related to these integrals satisfy $\frac{s}{\kappa} = c^2$.

1.6   Quaternions and Rotations in Three Dimensions

Rotations and translations of a material body in three dimensions are of major importance in many applications including Robotics. The first solution for this problem was made Euler who showed that any such three-dimensional rotation can be obtained as a superposition of three two-dimensional rotations in a fixed coordinate system. Since then, several variations and improvements (for specialized applications) of this algorithm appeared in the literature.

Another elegant solution to this problem requires the use of Quaternions.

In this section, we describe both of these approaches.

1.6.1   Euler's Angles

Two-dimensional rotations by a angle α around the z-axis of a fixed coordinate system in three dimensions is given by

$$R_z(\alpha) = \begin{pmatrix} \cos(\alpha) & \sin(\alpha) & 0 \\ -sin(\alpha) & \cos(\alpha) & 0 \\ 0 & 0 & 1 \end{pmatrix}. \tag{1.13}$$

Similarly rotations with angles β and γ around the x and y axis, respectively, are given by

$$R_x(\beta) = \begin{pmatrix} 1 & 0 & 0 \\ \cos(\beta) & \sin(\beta) & 0 \\ -sin(\beta) & \cos(\beta) & 0 \end{pmatrix}, \tag{1.14}$$

$$R_y(\gamma) = \begin{pmatrix} \cos(\gamma) & \sin(\gamma) & 0 \\ 0 & 1 & 0 \\ -sin(\gamma) & \cos(\gamma) & 0 \end{pmatrix}. \tag{1.15}$$

Suppose that we are given a (rigid) rotation (with no translation) in three dimensions. The original coordinate axes (x,y,z) are transferred then as a result of the rotation to (X,Y,Z). We want to represent this rotation To this end, define

$$\mathbf{N} = \mathbf{z} \times \mathbf{Z},$$

where **z** and **Z** are the unit vectors along the z and Z axes of the corresponding coordinates. Let

- ϕ to be the angle from the x axis to **N**.
- θ to be the angle from the z axis to the Z axis.
- ψ to be the angle from **N** to the X axis.

These angles are referred to as "Euler's Angles".

The matrix that represents this rotation from the coordinate system (x, y, z) to (X, Y, Z) is

$$R = R_z(\psi) R_x(\theta) R_z(\phi).$$

(This representation is not unique.)

1.6.2 Quaternions

All of us are familiar with the real and complex number systems. However, there are other number systems with "non-commutative unit numbers". The quaternion number system which was introduced by Hamilton 1843 is "based" on four "basis vectors" **1**, **i**, **j**, **k**. A general quaternion number is of the form

$$\mathbf{q} = a * \mathbf{1} + b * \mathbf{i} + c * \mathbf{j} + d * \mathbf{k},$$

where a, b, c, and d are real numbers.

The basis vectors satisfy the following relations:

$$(\mathbf{1})^2 = 1, \ (\mathbf{i})^2 = (\mathbf{j})^2 = (\mathbf{k})^2 = -1,$$

and

$$\mathbf{ijk} = -1.$$

A quaternion number **q** can be represented by a 2×2 matrix with complex numbers as follows:

$$q = \begin{pmatrix} a + bi & c + di \\ -c + di & a - bi \end{pmatrix}, \tag{1.16}$$

(Here "i" is the complex number i). In expanded form, this representation can be rewritten as

$$\mathbf{q} = a * \sigma_0 + b * \sigma_1 + c * \sigma_2 + d * \sigma_3,$$

where

$$\sigma_0 = \begin{pmatrix} 1 & 0 \\ 0 & 1 \end{pmatrix}, \quad \sigma_1 = \begin{pmatrix} i & 0 \\ 0 & -i \end{pmatrix}, \tag{1.17}$$

$$\sigma_2 = \begin{pmatrix} 0 & 1 \\ -1 & 0 \end{pmatrix}, \quad \sigma_3 = \begin{pmatrix} 0 & i \\ i & 0 \end{pmatrix}. \tag{1.18}$$

Thus σ_μ, $\mu = 0, 1, 2, 3$ are the "familiar" Pauli matrices.

1.6.2.1 Quaternions and Rotations in R^3

A complex number $z = re^{i\varphi}$ represents a point $p = (r\cos(\varphi), r\sin(\varphi))$ in two dimensions. We can then rotate p by an angle θ by multiplying it by $e^{i\theta}$. Thus $p' = re^{i(\varphi+\theta)}$. It follows then that in two dimensions, rotations can be represented as multiplication by a complex number of length 1. By analogy, we want to show that rotating a point (or a vector) in $3D$ might correspond to a multiplication by a unit quaternion that can be represented by an exponential.

To see how this can be done consider a vector $\mathbf{r} = (x, y, z)$. A rotation of this vector by an angle θ around the z axis is given by

$$\mathbf{r}' = R_z(\theta)\mathbf{r}, \tag{1.19}$$

where $\mathbf{r}'$ represents the rotated vector.

To convert this equation into "quaternion form", we represent the vector $\mathbf{r}$ as

$$\mathbf{q} = 0 * 1 + x * \mathbf{i} + y * \mathbf{j} + z * \mathbf{k}, \tag{1.20}$$

or equivalently as

$$\mathbf{q} = 0 * \sigma_0 + x * \sigma_1 + y * \sigma_2 + z * \sigma_3. \tag{1.21}$$

we now claim that the rotation in (1.19) can be rewritten in quaternion notation as

$$\mathbf{q}' = e^{-(\theta\sigma_3)/2}\mathbf{q}e^{(\theta\sigma_3)/2}, \tag{1.22}$$

to prove this statement note that the powers of the matrix σ_3 satisfy the following recursion

$$\sigma_3^2 = -I, \; \sigma_3^3 = -\sigma_3, \; \sigma_3^4 = I, \ldots,$$

(where I is unit two-dimensional matrix or σ_0). Hence

$$e^{\alpha * \sigma_3} = \cos(\alpha)\sigma_0 + \sin(\alpha)\sigma_3.$$

Therefore, (1.22) yields

$$\mathbf{q}' = [\cos(\theta/2)\sigma_0 - \sin(\theta/2)\sigma_3][x\sigma_1 + y\sigma_2 + z\sigma_3][\cos(\theta/2)\sigma_0 + \sin(\theta/2)\sigma_3] \tag{1.23}$$

$$= (x\cos(\theta) + y\sin(\theta))\sigma_1 + (-x\sin(\theta) + y\cos(\theta))\sigma_2 + z\sigma_3.$$

For rotations around other directions, one has to change σ_3 in (1.22) into $a * \sigma_1 + b * \sigma_2 + c * \sigma_3$ where $a^2 + b^2 + c^2 = 1$ to rotate around the axis represented by this unit quaternion.

Appendix 1A: The Catenary

The equation of the Caternary represents a model for the shape of a hanging chain or a rope.

To derive a model for the shape of the chain, we let the tension in the chain at s be denoted by $\mathbf{T} = \mathbf{T}(s)$. Since the chain is at rest, we have by Newton's second law

$$\mathbf{T}(s + \Delta s) - \mathbf{T}(s) + \mathbf{F}(s)\Delta s = 0,$$

where $\mathbf{F}(s)$ is the external force acting on the chain per unit length. From this equation we obtain in the limit $\Delta s \to 0$ that

$$\frac{d\mathbf{T}}{ds} = -\mathbf{F}(s).$$

If the only external force acting on the chain is the force of gravity and the chain is in the x-z plane, then $\mathbf{F} = (0, -\rho g)$ where ρ the mass density of the chain per unit length and g is the acceleration of gravity. Therefore,

$$\frac{d\mathbf{T}}{ds} = (0, \rho g),$$

which implies that

$$\mathbf{T} = (c, \rho g s),$$

where c is a constant. We now assume that $\mathbf{T}$ is tangent to the curve generated by the chain and let $\mathbf{u}$ is the unit tangent to the chain. By Newton's third law we must have $\mathbf{T} = T\mathbf{u}$ where T is constant (otherwise the chain is not static). Thus,

$$T\mathbf{u} = T(\frac{dx}{ds}, \frac{dy}{ds}) = (c, \rho g s).$$

This implies that

$$\frac{dy}{dx} = \frac{dy/ds}{dx/ds} = \frac{\rho g s}{c} = \frac{s}{b}, \quad b = \frac{\rho g}{c}.$$

Hence

$$\frac{dx}{ds} = \frac{1}{\frac{ds}{dx}} = \frac{1}{\sqrt{1 + (\frac{dy}{dx})^2}} = \frac{b}{\sqrt{b^2 + s^2}}.$$

Similarly

$$\frac{dy}{ds} = \frac{dy}{dx}\frac{dx}{ds} = \frac{s}{\sqrt{b^2 + s^2}}.$$

Integrating we obtain (up to constants which we set to zero)

$$x = b \operatorname{arcsinh}(\frac{s}{b}), \quad y = \sqrt{b^2 + s^2}.$$

Therefore $\frac{s}{b} = \sinh(\frac{x}{b})$ i.e. $s = b \sinh(\frac{x}{b})$. Substituting this in the expression for y leads to

$$y = b\sqrt{1 + \sinh^2(\frac{x}{b})} = b \cosh(\frac{x}{b}).$$

Finally, the equation for the chain curve is

$$\mathbf{x} = (x, b \cosh(\frac{x}{b})).$$

Appendix 1B: Astrobiology and Curves in R^3

It was discovered around 1953 that the DNA in living organisms on Earth has a double Helix structure. Why nature chose this geometrical structure as a building block of life on Earth remains unanswered. This raises the possibility that other life forms might use as building blocks molecules with different geometrical structure. In this appendix, we explore some of the mathematical aspects of this problem in order to gain some insights into this important issue and make "educated guesses" about other possible forms of life in the Universe.

1B.1 Introduction

A major discovery about life as we know it was made in around 1953 (Watson & Crick 1953). At that time, it was discovered that the "basis" of all life on Earth revolves around the "DNA molecule", which has a double helix structure. This discovery gave rise to a new research discipline about "On the Origin of Life" (Eschenmoser 2007, Greenberg et al. 1991, Mariscal et al. 2019). This discipline attempts to explore the chemical and biological pathways that gave rise to the DNA molecule. It has led to many significant insights about this process. A comprehensive list of research papers on this subject can be found in (Mariscal et al. 2019). However, whether there exist other processes which might lead to "Life" with molecules of other chemical and/or geometrical structures remain as an open question. Thus, as far as we know, no one was able to prove so far that the chemical and biological processes that give rise to the DNA molecule have **one unique result**. (Especially since these processes are subject to a myriad of random influences). Naturally this raises the question whether alternative structures are possible and can support other forms of "Life". In fact in a recent article some authors (Petkowski et al. 2025) speculated that another molecule can also support different form of life.

In this presentationm we explore this issue from a mathematical point of view by reviewing the special geometrical characteristic of the helix and highlighting those geometrical structuresm which come close to those of the DNA molecule (Babaarslan & Yayli 2013, Monterde 2007, Deshmukh et al. 2019). We can provide no biological or chemical evidence for the existence of other forms of Life which have these geometrical structures. Howeverm the fact that these structures might "deviate marginally" from the helix structure of the DNA molecule make it feasible to speculate that such life forms might actually exist. As humanity embarks on the exploration of other objects in the solar system, we must be prepared to face the unexpected surprises that the unknown has in its folds for us.

Any attempt to guess the geometrical structures of molecules, other than the Helix, that can support a form of "LIFE" (viz. self replicating) is nothing but pure speculation. However, we might be able to make "educated guesses" guided by what is known already.

1B.2 Other Forms of Life

As we showed above, a helix is characterized by the fact that $\frac{\tau(s)}{\kappa(s)}$ is a constant(Lancert 1806). This relationship can be rewritten

$$\tau(s) - \lambda\kappa(s) = 0, \tag{1B. 1}$$

where λ is a constant. That is there is a linear relationship between $\tau(s)$ and $\kappa(s)$. The most natural generalization of this relation is to replace it by a polynomial relationship viz.

$$P(\tau(s)) = \lambda\kappa(s). \tag{1B. 2}$$

where P is a polynomial. The simplest example for such a relationship is to replace (1B. 1) by

$$a\tau(s) + b\kappa(s) = 1, \tag{1B. 3}$$

where a, b are constants and $\kappa(s)$ is not constant. This relationship characterizes Bertrand curves (Monterde 2007, Bertrand 1850, Babaarslan & Yayli 2013). An explicit example of such a curve (up to integration) is given by (Monterde 2007, Teixeira 1995)

$$\frac{d\mathbf{x}}{dt} = \left(\frac{t}{t^2 + 2} + \frac{(t^2 + 2)^{1/2}}{(t^2 + 1)^{3/2}}\right)(\sin t, \cos t, t). \tag{1B. 4}$$

The curvature and torsion of this curve are

$$\kappa = \frac{(t^2+2)^{3/2}}{t(t^2+1)^{3/2}+(t^2+2)^{3/2}}, \quad \tau = \frac{t(t^2+1)^{3/2}}{t(t^2+1)^{3/2}+(t^2+2)^{3/2}}. \qquad (1\text{B}.\,5)$$

and we have $\tau + \kappa = 1$. (Monterde 2007, Monterde 2009)

Another way to construct Bertrand curves in three dimensions is to consider those with constant curvature κ and non-constant torsion. Equation (1B. 3) is satisfied then if we let $a = 0$ and $b = \frac{1}{\kappa}$. These curves are referred to as Salkowski curves (Salkowski 1909, Monterde 2009). An example of such a curve with $\kappa = 1$ and $\tau = s/10$ is shown in Figure 1.5. (Biologically, one might attempt to achieve this geometrical molecular form by deformation of the helical geometry of the DNA molecule).

Another general algorithm to construct Bertrand curves was given by (Izumiya and Takeuchi 2002) (Theorem 4.1 on p. 102 of this reference).

In this algorithm, one starts with a spherical curve (viz. a curve on a sphere) $\mathbf{x}(s)$, which is parameterized by its arc length s,

$$\mathbf{x}(s) = (x(s), y(s), z(s)), \quad x(s)^2 + y(s)^2 + z(s)^2 = R^2, \qquad (1\text{B}.\,6)$$

where R is the sphere radius. Let

$$\mathbf{T}(s) = \left(\frac{dx}{ds}, \frac{dx}{ds}, \frac{dx}{ds}\right), \qquad (1\text{B}.\,7)$$

be the tangent to this curve. Define

$$\mathbf{V}(s) = \mathbf{x}(s) \times \mathbf{T}(s), \qquad (1\text{B}.\,8)$$

FIGURE 1.5
Hanging rope.

Bertrand Curve with $\kappa=1$ $\tau=$s/10

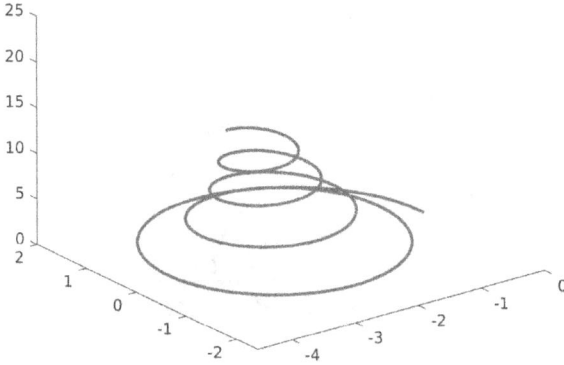

FIGURE 1.6
An example of Bertrand curve.

($\mathbf{V}(s)$ is the vector product of $\mathbf{x}(s)$ and $\mathbf{T}(s)$). Then,

$$A \int \mathbf{x}(s)ds + A \cot \theta \int \mathbf{V}(s)ds, \qquad (1\text{B. }9)$$

(where A and θ are constants) is a Bertrand curve (Figure 1.6).

1B.3 Two-Dimensional Life Forms

Is it possible that some "Life" forms might have almost two-dimensional structure (that is flat). As in three dimensions, we look for geometrical structures that are based on two-dimensional curves with some distinctive features.

As an example, we consider the "Logarithmic Spiral" (Figure 1.7) whose equation in polar coordinates (viz. $x = r \cos \theta$, $y = r \sin \theta$) is

$$r(\theta) = e^{a\theta}, \quad a > 0. \qquad (1\text{B. }10)$$

That is

$$\mathbf{x}(\theta) = (e^{a\theta} \cos \theta, e^{a\theta} \sin \theta, 0). \qquad (1\text{B. }11)$$

For this spiral, it is easy to show that the arc length of the curve (measured from the origin which corresponds to $\theta = -\infty$) is

$$s(\theta) = \frac{\sqrt{a^2 + 1}}{a} e^{a\theta}. \qquad (1\text{B. }12)$$

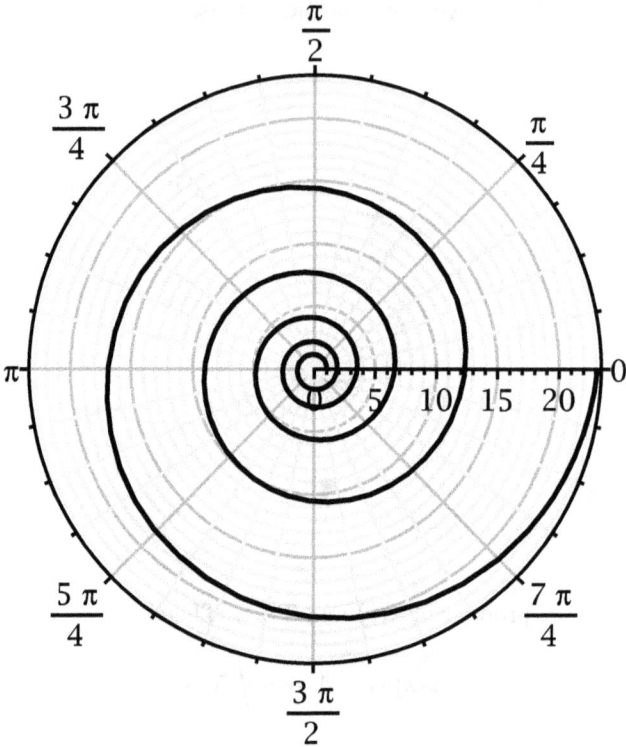

FIGURE 1.7
Logarithmic spiral.

and therefore $s(\theta)$ is proportional to $r(\theta)$. The unit length tangent vector to this curve is

$$\mathbf{T} = \frac{1}{\sqrt{a^2 + 1}}(a\cos\theta - \sin\theta, a\sin\theta + \cos\theta, 0). \qquad (1\text{B. }13)$$

Hence, the angle ϕ between $\mathbf{T}$ and $\mathbf{x}$ satisfies,

$$\cos\phi = \frac{a}{\sqrt{a^2 + 1}}, \qquad (1\text{B. }14)$$

which implies that ϕ is constant. Finally this curve has a curvature

$$\kappa = \frac{e^{-a\theta}}{\sqrt{a^2 + 1}}. \qquad (1\text{B. }15)$$

Thus κ goes to zero as $\theta \to \infty$.

Due to all these properties, Jacob Bernoulli dubbed this curve as "Spira mirabilis" (Latin for "miraculous spiral") (Bernoulli 1713, Struik 1988). In fact, some shells on Earth display this geometrical structure. For example, the chambers of the nautilus shell are arranged approximately in a logarithmic spiral. Similarly, the arms of spiral galaxies and cyclones on Earth display (in many cases) the approximate shape of a logarithmic spiral (Figure 1.7).

1B.4 Conclusions

In this appendix, we asked why nature chose a helix as a building block of life. To answer this question from a mathematical point of view, we quoted Lancert's theorem that shows that helices are characterized geometrically by the fact that the ratio of the curve torsion to its curvature is constant. Based on this observation, we speculated that other (possible) life forms might satisfy different (but closely related) mathematical relationships between torsion and curvature. It might argued, justifiably, that we do not have any experimental evidence to backup our speculations. However, history shows that there were cases in the past where theory preceded experiment and stimulated research in new venues. This presentation also demonstrates the fact that some mathematical topics that were considered once as abstract might turn out to have important applications in "real life". In fact, in view of the importance of this topic, NASA has a program on "Astrobiology" (https://astrobiology.nasa.gov/. To join live presentations, mail to https://hq-astrobiology@mail.nasa.gov)

Acknowledgment

The author is indebted to Professors. J. Monterde and Y. Yayli for their helpful input in the preparation of this appendix.

Appendix 1C: Maple Programs

With this chapter, we provide four maple programs that can be used to compute the curvature and torsion of the following curves in R^3.

- circular-helix.mw: elliptic Helix

- catenary.mw: catenary curve

- Folium.mw: FOlium of Descartes

- spherical-curve.mw: a curve on the sphere.

Each of these programs can be easily modified to compute the curvature and torsion of other curves.

Exercises

1. The logarithmic spiral is defined in polar coordinates by $r(t) = exp(t)$, $\theta = at$ (where a is a constant).

   - Plot this curve for $t \in [0, \frac{6\pi}{a}]$.
   - Find the arc length s along this curve and express the spiral equation in terms of this parameter.
   - Show that the position vector has a constant angle with the tangent vector.

2. Derive the explicit form of Frenet equations for the ellipse $\frac{x^2}{a^2} + \frac{y^2}{b^2} = 1$. What happens when $a = b$ (circle of radius a.

3. Darboux rotation vector is defined as,

$$\mathbf{D} = \tau\mathbf{T} + \kappa\mathbf{b}.$$

   Use Frenet formulas to compute $\mathbf{D} \times \mathbf{T}$, $\mathbf{D} \times \mathbf{p}$, $\mathbf{D} \times \mathbf{b}$.

4. Use (3.22) and (1.7) to compute the curvature of the ellipse and hyperbola.

5. Show that for plane curve $y(x)$ i.e $\mathbf{x} = (x, y(x), 0)$

$$\kappa = \frac{|y''|}{(1 + (y')^2)^{3/2}}$$

6. Compute the curvature and torsion of the Caternary.

7. Compute the curvature and torsion of Fermat Spiral $r = A\theta^{(}1/2)$

8. Compute the curvature and torsion of the hyperbolic spiral $r = A/\theta$.

9. compute (and verify) the quaternion formula (1.22) for the transformation $\mathbf{r}' = R_y(\theta)\mathbf{r}$. Note that you will have to change σ_3 to σ_2 in (1.22).

Further Readings

M. Babaarslan, Y. Yayli (2013), On Helices and Bertrand curves in Euclidean 3-space, *Math. Comput. Appl.*, 18(1), 1–11.

M. Barros (1997), General helices and a theorem of Lancret, *Proc. Am. Math. Soc.*, 125, 1503–1509.

J. Bernoulli (1713), *Ars Conjectandi*, Thurnisiorum, fratrum, Basileae.

J. Bertrand (1850), Memmoire sur la theorie des courbes à double courbure, *Comptes Rendus*, 31, 332–350.

S. Deshmukh, A. Alghanemi, R. T. Farouki (2019), Space curves defined by curvature-torsion relations and associated helices, *Filomat*, 33(15), 4951–4966. https://doi.org/10.2298/FIL1915951D.

A. Eschenmoser (2007), The search for the chemistry of life's origin, *Tetrahedron*, 63, 12821–12844. https://doi.org/10.1016/j.tet.2007.10.01.

J. M. Greenberg, C. X. Mendoza-Gomez, V. Pironello (eds.) (1991), The chemistry of life's origins, *Proceedings of the NATO Advanced Study Institute and 2nd International School of Space Chemistry*, Erice, Sicily, Italy 20–30 October 1991, ISBN 978-94-010-4856-9.

S. Izumiya, N. Takeuchi (2002), Generic properties of helices and Bertrand curves, *J. Geometry.*, 74, 97–109.

E. Kreyszig (1991), *Differential Geometry*, Dover, New York.

M. A. Lancret (1806), Mémoire sur les courbes à double courbure. *Mémoires présentés à l'Institut*, 1, 416–454.

C. Mariscal, A. Barahona, N. Aubert-Kato (2019), Hidden concepts in the history and philosophy of origins-of-life studies: A workshop report, *Orig. Life Evol. Biosph.*, https://doi.org/10.1007/s11084-019-09580-x.

J. Monterde (2007), Curves with constant curvature ratios, *Boletín de la Sociedad Matemática Mexicana 3a*, 13(1), 177–186.

J. Monterde (2009), Salkowski curves revisited: A family of curves with constant curvature and non-constant torsion, *Comput. Aided Geom. Des.*, 26(3), 271–278.

J.J. Petkowski et al.(2025), Astrobiological implications of the stability, *Sci. Adv.*, 11, eadr0006.

E. Salkowski (1909), Zur Transformation von Raumkurven, *Mathematische Annalen*, 66(4), 517–557.

D. J. Struik (1988), *Lectures on Classical Differential Geometry*, Dover, New-York.

F. G. Teixeira (1995), *Traité des courbes spéciales remarquables planes et gauches*, t.2, p. 447. Chelsea Pub. Co, Bronx, N.Y.

J. D. Watson, F. Crick (1953), Molecular structures of nucleic acids, *Nature*, 171, 737–738.

2

Introduction to Classical Riemannian Geometry

2.1 Surfaces in R^3

2.1.1 Some Important Surfaces in R^3

Definition 2.1: A Quadratic Surfaces is a surface in R^3 that satisfies an equation of the form

$$Ax^2 + By^2 + Cz^2 + Dx + Ey + Fz = H,$$

where A, B, C, D, E, F, H are constants. By translations and rotation, these surfaces can be brought into a "canonical form" and classified according to their "geometric characteristics". The following is an enumeration of these families:

1. Ellipsoids

$$\frac{x^2}{a^2} + \frac{y^2}{b^2} + \frac{z^2}{c^2} = 1.$$

2. Elliptic Parabolods

$$\frac{x^2}{a^2} + \frac{y^2}{b^2} = \frac{z}{c}.$$

3. Cones

$$\frac{x^2}{a^2} + \frac{y^2}{b^2} = \frac{z^2}{c^2}.$$

4. Hyperboloids of One Sheet

$$\frac{x^2}{a^2} + \frac{y^2}{b^2} - \frac{z^2}{c^2} = 1.$$

5. Hyperboloids of Two Sheets

$$-\frac{x^2}{a^2} - \frac{y^2}{b^2} + \frac{z^2}{c^2} = 1.$$

DOI: 10.1201/9781003587422-2

6. Hyperbolic Paraboloids (Saddles)

$$\frac{x^2}{a^2} - \frac{y^2}{b^2} = \frac{z}{c}, \quad c > 0.$$

Definition 2.2: A surface in R^3 is called a **Surface of Rotation** if it can be obtained by rotating a regular plane curve in the x-z plane around the z-axis.

If the curve in the x-z plane is

$$\mathbf{x}(t) = (r(t), 0, h(t)),$$

then the equation of the surface generated by rotating this curve around the z-axis is

$$\mathbf{x}(t, \phi) = (r(t) \cos \phi, r(t) \sin \phi, h(t)).$$

For example, a sphere is obtained by rotating a circle whose center is at the origin around the z-axis. Similarly, a cylinder is obtained by rotating a straight line in the x-z plane (not passing through the origin) around the z-axis.

2.1.1.1 Torus

The Torus is a donut-shaped surface that is obtained by rotating a circle whose center is at $(a, 0, 0)$ $(a \neq 0)$ and (constant) radius $r < a$ around the z-axis. The equation of the circle in x-z plane is

$$\mathbf{x}(\theta) = (a + r \cos \theta, 0, r \sin \theta), \quad a > r, \quad 0 \leq \theta \leq 2\pi.$$

Therefore, the equation for the surface that is generated by rotating this curve around the z-axis is (Figures 2.1–2.3)

$$\mathbf{x}(\theta, \phi) = ((a + r \cos \theta) \cos \phi, (a + r \cos \theta) \sin \phi, r \sin \theta), \quad 0 \leq \phi \leq 2\pi, \ 0 \leq \theta \leq 2\pi.$$

Definition 2.3: A surface in R^3 is called a **Ruled Surface** if its equation can be written as

$$\mathbf{x}(u, v) = \mathbf{c}(u) + v\mathbf{r}(u), \quad \frac{d\mathbf{r}}{du} \neq 0.$$

Usually $\mathbf{r}$ is taken as a unit vector.

Example 2.1: Show that the hyperboloid of one sheet

$$x^2 + y^2 - z^2 = 1,$$

is a ruled surface.

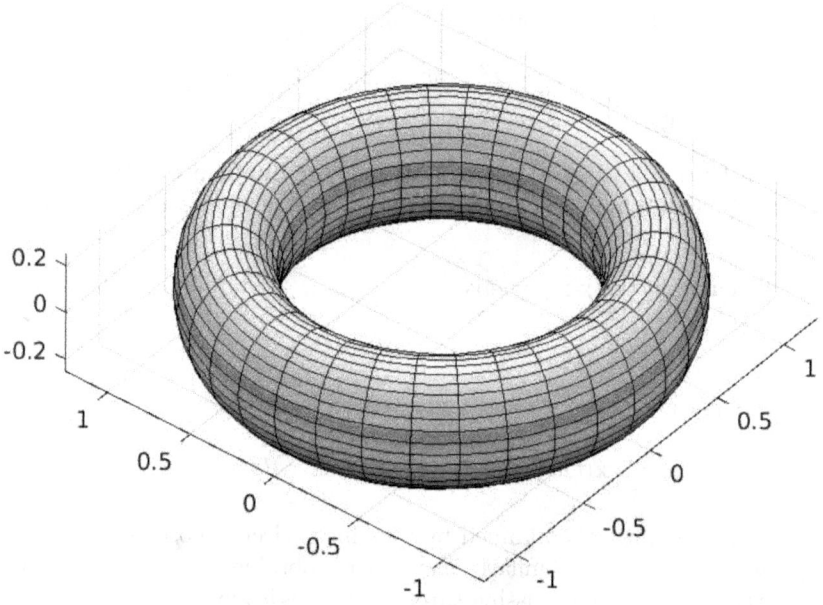

FIGURE 2.1
A torus.

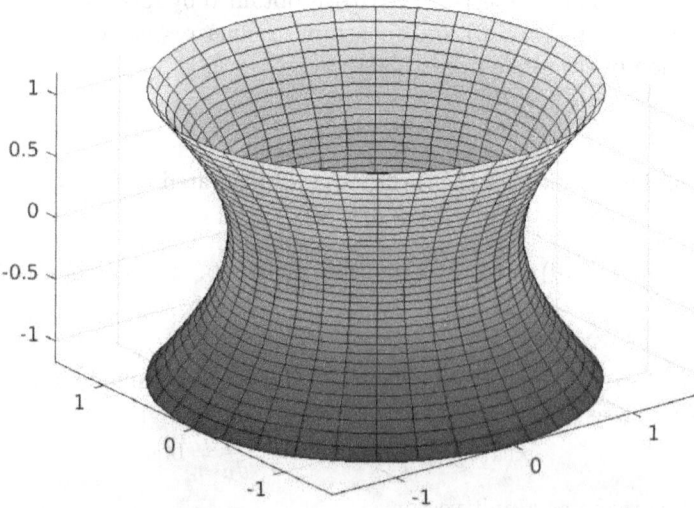

FIGURE 2.2
Hyperboloid of one sheet.

Solution: First, we observe that a straight line on the surface exists in the form $\mathbf{w} = (1, u, u)$. Now we want to represent the equation of this hyperboloid in the form

$$\mathbf{x}(u, v) = \mathbf{w}(u) + v(f(u), g(u), h(u)).$$

Substituting this in the original (Cartesian) equation for the surface yields

$$(1 + vf(u))^2 + (u + vg(u))^2 - (u + vh(u))^2 = 1,$$

which leads to

$$(f^2 + g^2 - h^2)v^2 + 2(f + gu - hu)v = 0.$$

Since v is a variable this implies that

$$f^2 + g^2 - h^2 = 0, \quad f + gu - hu = 0.$$

One possible solution to these equations is

$$f = 2u, \quad g = u^2 - 1, \quad h = u^2 + 1.$$

We observe that a second solution of this system in the form

$$f = 0, \quad g = 1, \quad h = 1,$$

is not admissible since then

$$\mathbf{x}(u, v) = (1, u + v, u + v),$$

is an equation of a line.

We note that actually this hyperboloid is "doubly ruled" (see picture below). The equations of these two sets of lines on the surface of the hyperboloid $\left(\frac{x}{a}\right)^2 + \left(\frac{y}{b}\right)^2 - \left(\frac{z}{c}\right)^2 = 1$ are

$$\mathbf{x} = (a(\cos\theta + t\sin\theta), b(\sin\theta - t\cos\theta), \pm ct), \quad \theta \in [0, 2\pi], \quad t \in R.$$

Example 2.2: Show the "Mobius strip" is a ruled surface. The equation of this surface is

$$\mathbf{x}(u, v) = \left(\sin u(1 + v\sin(\tfrac{u}{2})), \cos u(1 + v\sin(\tfrac{u}{2})), v\cos(\tfrac{u}{2})\right),$$
$$u \in R, \quad v \in [-1, 1].$$

This can be rewritten as

$$\mathbf{x}(u, v) = (\sin u, \cos u, 0) + v\left(\sin u \sin(\tfrac{u}{2}), \cos u \sin(\tfrac{u}{2}), \cos(\tfrac{u}{2})\right).$$

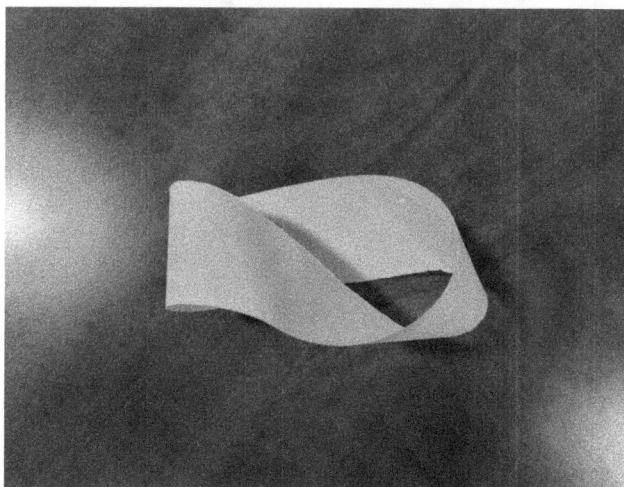

FIGURE 2.3
A Mobius strip.

2.1.2 Surfaces in R^3

Let (x, y, z) be a coordinate system in R^3 (Euclidean three- dimensional space). Let $\mathbf{x}(u_1, u_2)$ be a smooth function on a bounded set $U \subset R^2$ to R^3. (That is $\mathbf{x} : U \subset R^2 \to R^3$).

Definition 2.4: We shall say that

$$\mathbf{x}(u_1, u_2) = (x(u_1, u_2), y(u_1, u_2), z(u_1, u_2)).$$

is a surface (manifold) S in R^3 if $\mathbf{x}$ is one-to-one and the Jacobian of $\mathbf{x}$

$$J = \begin{pmatrix} \frac{\partial x}{\partial u_1} & \frac{\partial x}{\partial u_2} \\ \frac{\partial y}{\partial u_1} & \frac{\partial y}{\partial u_2} \\ \frac{\partial z}{\partial u_1} & \frac{\partial z}{\partial u_1} \end{pmatrix},$$

at each point of B is of rank two.

Example 2.3: The function

$$\mathbf{x}(u_1, u_2) = (u_1 + u_2, (u_1 + u_2)^2, (u_1 + u_2)^4),$$

does NOT represent a surface in R^3 since at each point its Jacobian has rank 1. In fact if we let $t = u_1 + u_2$ then it becomes obvious that $\mathbf{x}$ represents the curve (t, t^2, t^4).

Remark 2.1: In some cases, J has rank 2 on all of B except for a finite number of points. In these cases, we shall refer to these points as *singular points* of the surface with respect to the given representation.

2. In calculus one is used to represent a surface in the form $z = F(x, y)$, one can obtain from this representation the one above by letting $u_1 = x$, $u_2 = y$ and the (vector) representation of the surface is $\mathbf{x} = (u_1, u_2, F(u_1, u_2))$ whose Jacobian is obviously of rank 2.

Example 2.4: The x-y plane in R^3 has the following two representations:

- $\mathbf{x} = (u_1, u_2, 0)$

- $\mathbf{x} = (u_1 \cos u_2, u_1 \sin u_2, 0)$.

For both representations, the Jacobian matrix has rank two in general. However, in the second representation, the point $(0, 0, 0)$ is a singular point.

Example 2.5: The sphere $x^2 + y^2 + z^2 = R^2$ have a "Cartesian representation" in terms of two hemispheres $z = \pm\sqrt{R^2 - x^2 - y^2}$. It has also a representation in terms of spherical coordinates

$$\mathbf{x} = (R \sin u_1 \cos u_2, R \sin u_1 \sin u_2, R \cos u_1).$$

In this representation, the two poles are singular points.

Example 2.6: A circular cone has the following Cartesian representation

$$\mathbf{x} = (u_1, u_2, \pm R\sqrt{u_1^2 + u_2^2}),$$

which utilizes two sheets. Another representation is

$$\mathbf{x} = (u_1 \cos u_2, u_1 \sin u_2, R u_1).$$

In this representation, the curves $u_1 = $ constant are circles. The cone apex $u_1 = 0$ is a singular point.

2.2 Tangent Planes

A parametric representation of a curve on a surface S is a mapping $t \to (u_1(t), u_2(t))$. That is, such a curve on S is given by $\mathbf{x}(u_1(t), u_2(t))$

Example 2.7: A circular cylinder is given by

$$\mathbf{x} = (r \cos u_2, r \sin u_2, u_1).$$

The helix $\mathbf{x} = (r \cos t, r \sin t, at)$ is a curve on this cylinder with $u_1 = at$, $u_2 = t$.

Definition 2.5: A family of curves on S which depend smoothly on a parameter α is called a "one parameter family of curves on S". Two such (distinct) one parameter families of curves is called a net on S if at each point p of S only one curve of each family pass through p and these curves have different tangents at p.

Example 2.8: Let $\mathbf{x}$(u,v) be a surface in R^3. The coordinate curves

$$\mathbf{x}(u, b) = (x(u, b), y(u, b), z(u, b)),$$

and

$$\mathbf{x}(a, v) = (x(a, v), y(a, v), z(a, v)),$$

where a, b are constants form a net on S since each point of S is represented by unique values of u, v and the tangents to these curves at the point $(u, v) = (a, b)$ are independent due to the requirement that the Jacobian of the surface has rank 2.

The tangent vector to a curve $(u(t), v(t))$ on S is given by

$$\frac{d\mathbf{x}}{dt} = (\frac{dx}{dt}, \frac{dy}{dt}, \frac{dz}{dt}) = \frac{\partial \mathbf{x}}{\partial u}\frac{du}{dt} + \frac{\partial \mathbf{x}}{\partial v}\frac{dv}{dt} = \mathbf{x}_u\frac{du}{dt} + \mathbf{x}_v\frac{dv}{dt}. \qquad (2.1)$$

We infer from this formula that the tangent vector of any curve on S which passes through a point p is a linear combination of the vectors $\mathbf{x}_u(p)$, $\mathbf{x}_v(p)$.

We define therefore the tangent plane at a point p as

$$\mathbf{y} = \mathbf{x}(p) + \alpha\mathbf{x}_u(p) + \beta\mathbf{x}_v(p).$$

In other words, the vector $\mathbf{y} - \mathbf{x}(p)$ is a linear combination of the vectors $\mathbf{x}_u(p)$, $\mathbf{x}_v(p)$ and, therefore, for any point $\mathbf{y}$ on the tangent plane at p the determinant

$$\det(\mathbf{y} - \mathbf{x}(p), \mathbf{x}_u(p), \mathbf{x}_v(p)) = 0.$$

2.3   First Fundamental Form of a Surface

Let $\mathbf{x}(u_1, u_2)$ be a surface in R^3. The infinitesimal arc length of a curve on this surface is

$$ds^2 = \mathbf{dx}(u_1(t), u_2(t)) \cdot \mathbf{dx}(u_1(t), u_2(t)).$$

From Eq. (3.3) we have for a curve on a surface

$$\mathbf{dx} = \mathbf{x}_{u_1}du_1 + \mathbf{x}_{u_2}du_2.$$

Hence

$$ds^2 = g_{11}du_1^2 + 2g_{12}du_1du_2 + g_{22}du_2^2 = \sum_{i,j} g_{ij}du_idu_j, \qquad (2.2)$$

where

$$g_{11} = \mathbf{x}_{u_1}^2, \quad g_{12} = g_{21} = \mathbf{x}_{u_1} \cdot \mathbf{x}_{u_2}, \quad g_{22} = \mathbf{x}_{u_2}^2.$$

The quadratic form (2.2) is called "the first fundamental form of the surface", and the coefficients g_{ij} are referred to as the "coefficients of the metric tensor". It should be observed that these coefficients are dependent on the choice of the parametric representation of the surface.

Example 2.9: A sphere of radius R is represented in spherical (geographic) coordinates as

$$\mathbf{x} = (R\cos u_1 \cos u_2, R\cos u_1 \sin u_2, R\sin u_1).$$

It follows then that,

$$\begin{aligned} \mathbf{x}_{u_1} &= (-R\sin u_1 \cos u_2, -R\sin u_1 \sin u_2, R\cos u_1), \\ \mathbf{x}_{u_2} &= (-R\cos u_1 \sin u_2, R\cos u_1 \cos u_2, 0). \end{aligned}$$

Hence

$$g_{11} = R^2, \quad g_{12} = 0, \quad g_{22} = R^2 \cos^2 u_1.$$

Therefore,

$$ds^2 = R^2(du_1)^2 + R^2 \cos^2 u_1 (du_2)^2.$$

Example 2.10: The equation of an elliptic cylinder is

$$\mathbf{x} = (a\cos(u_1), b\sin(u_1), u_2), \quad u_1 \in [0, 2\pi], \quad u_2 \in R.$$

Therefore

$$\begin{aligned} d\mathbf{x} &= (-a\sin(u_1)du_1, b\cos(u_1)du_1, du_2) \qquad (2.3) \\ &= (-a\sin(u_1), b\cos(u_1), 0)du_1 + (0, 0, 1)du_2 \\ &= \mathbf{x}_{u_1} du_1 + \mathbf{x}_{u_2} du_2. \end{aligned}$$

Hence

$$g_{11} = \mathbf{x}_{u_1}\cdot\mathbf{x}_{u_1} = a^2 \sin^2(u_1) + b^2 \cos^2(u_1), \quad g_{12} = \mathbf{x}_{u_1}\cdot\mathbf{x}_{u_2} = 0, \quad g_{22} = \mathbf{x}_{u_2}\cdot\mathbf{x}_{u_2} = 1.$$

$$ds^2 = g_{11}du_1^2 + du_2^2.$$

Example 2.11: The equation of the torus is

$$\mathbf{x}(\theta, \phi) = ((a+r\cos\theta)\cos\phi, (a+r\cos\theta)\sin\phi, r\sin\theta), \quad 0 \le \phi \le 2\pi, \ 0 \le \theta \le 2\pi,$$

therefore

$$\mathbf{x}_\theta = (-r\sin\theta\cos\phi, -r\sin\theta\sin\phi, r\cos\theta),$$

$$\mathbf{x}_\phi = (-(a + r\cos\theta)\sin\phi, (a + r\cos\theta)\cos\phi, 0).$$

Hence,

$$g_{11} = r^2, \quad g_{12} = g_{21} = 0, \quad g_{22} = (a + r\cos\theta)^2.$$

Definition 2.6: A metric that is defined by a positive definite quadratic form, i.e., the eigenvalues of the matrix

$$g = \begin{pmatrix} g_{11} & g_{12} \\ g_{21} & g_{22} \end{pmatrix},$$

are positive is called a "Riemannian metric". The geometrical treatment of surfaces with Riemannian metric is referred to as "Riemannian Geometry". If this quadratic form is not positive definite, the metric is called "Pseudo-Riemannian metric". The corresponding manifolds (=surfaces) are referred to as "Pseudo-Riemannian spaces". Such manifolds play an important role in special and general relativity.

Example 2.12: In special relativity, space and time are considered as one manifold and ds^2 is

$$ds^2 = dx^2 + dy^2 + dz^2 - c^2 dt^2,$$

i.e $g_{11} = g_{22} = g_{33} = 1$, $g_{44} = -c^2$ and $g_{ij} = 0$ when $i \neq j$. It follows that one of the eigenvalues of the matrix g is negative. and the space-time manifold is a four-dimensional Pseudo-Riemannian space.

Remark 2.2:

1. The entries in the inverse of g will be denoted by g^{ij}.

$$g^{11} = \frac{g_{22}}{\det(g)}, \quad g^{22} = \frac{g_{11}}{\det(g)}, \quad g^{12} = g^{21} = \frac{-g_{12}}{\det(g)},$$

2. Under a (nonsingular) coordinate transformation $w_i = w_i(u_1, u_2)$, $i = 1, 2$ (i.e. the Jacobian of the transformation is nonsingular), we have

$$du_i = \sum_m \frac{\partial u_i}{\partial w_m} dw_m.$$

Therefore,

$$ds^2 = \sum_{ij} g_{i,j} du_i du_j = \sum_{ij} g_{ij} \sum_m \frac{\partial u_i}{\partial w_m} dw_m \sum_n \frac{\partial u_j}{\partial w_n} dw_n = \sum_{mn} \bar{g}_{mn} dw_m dw_n,$$

where

$$\bar{g}_{mn} = \sum_{ij} g_{ij} \frac{\partial u_i}{\partial w_m} \frac{\partial u_j}{\partial w_n}.$$

Similarly, by switching the roles of u_i and w_j, we have

$$g_{ij} = \sum_{mn} \bar{g}_{mn} \frac{\partial w_m}{\partial u_i} \frac{\partial w_n}{\partial u_j}.$$

2.4 Other Fundamental Forms

The normal at a point to a surface in R^3 is given by the cross product of the tangent vectors

$$\bar{\mathbf{N}} = \mathbf{x}_{u_1} \times \mathbf{x}_{u_2}.$$

The unit normal is

$$\mathbf{N} = \frac{\bar{\mathbf{N}}}{|\bar{\mathbf{N}}|}.$$

Definition 2.7: The mapping

$$\mathbf{N}(u_1, u_2) : U \to S^2 \subset R^3,$$

where S^2 is the unit sphere in R^3 is called the "Gauss Map"

Definition 2.8: A surface is **Orientable** if the mapping

$$\mathbf{x}(u_1, u_2) \to (\mathbf{x}(u_1, u_2), \mathbf{N}(u_1, u_2)),$$

is continuous.

Example 2.13: The sphere is orientable.

Example 2.14: The Mobius strip is not orientable

To show this observe that $\mathbf{x}(0,0) = \mathbf{x}(2\pi, 0)$. However $\mathbf{N}(0,0) = (0,1,0)$ while $\mathbf{N}(2\pi, 0) = (0, -1, 0)$. (See maple attachment).

We now define the second and third fundamental forms of a surface

Definition 2.9: The second fundamental form of a surface is

$$h_{ij} = -\frac{\partial \mathbf{N}}{\partial u_i} \cdot \frac{\partial \mathbf{x}}{\partial u_j}.$$

Definition 2.10: The third fundamental form of a surface is

$$e_{ij} = \frac{\partial \mathbf{N}}{\partial u_i} \cdot \frac{\partial \mathbf{N}}{\partial u_j}.$$

2.4.1 Gaussian and Mean Curvature

Let P be a point on a manifold S and consider all the curves on this manifold pass through this point. Each of these curves C shall have a certain curvature $\kappa(C, P)$ which (in general) will depend on the curve and the point P. The set of all these curvature shall have a maximum $\kappa_1(P)$ and a minimum $\kappa_2(P)$.

Definition 2.11: The product

$$K(P) = \kappa_1(P)\kappa_2(P),$$

is called the Gaussian curvature of the manifold at P. Similarly,

$$M(P) = \frac{\kappa_1(P) + \kappa_2(P)}{2},$$

is called the mean curvature of S at P.

Theorem 2.1: Let

$$H(P) = \det(h_{ij}) = h_{11}h_{22} - h_{12}h_{21}, \quad G(P) = \det(g_{ij}) = g_{11}g_{22} - g_{12}^2.$$

Then

$$K(P) = \frac{H(P)}{G(P)}.$$

However, Gauss was able to show that $K(P)$ can be expressed without referring to the second fundamental form.

2.4.1.1   Gauss Theorem Egregium

$K(P)$ depends only on the metric tensor and its first- and second-order derivatives.

This theorem implies that all geometrical properties of a two-dimensional manifold can be deduced from measurements done on the manifold itself without referring to an embedding of the manifold in a space of larger dimension. As a result, the second and third fundamental forms have limited use. However, they are still utilized in many books and applications to motivate and derive some results geometrically.

Example 2.15: Calculate the coefficients of the second fundamental form for the torus and the expression for its Gaussian curvature.

The equation of the torus is

$$\mathbf{x}(\theta, \phi) = ((a+r\cos\theta)\cos\phi, (a+r\cos\theta)\sin\phi, r\sin\theta), \quad 0 \le \phi \le 2\pi, \ 0 \le \theta \le 2\pi.$$

and its metric tensor is (see previous example)

$$g_{11} = r^2, \quad g_{12} = g_{21} = 0, \quad g_{22} = (a + r\cos\theta)^2.$$

The coefficients of the second fundamental form are then

$$h_{11} = -(a + r\cos\theta)^2\cos\theta, \quad h_{12} = h_{21} = 0, \quad h_{22} = -r.$$

Hence

$$K = \frac{\det(h_{ij})}{\det(g_{ij})} = \frac{\cos\theta}{r(a + r\cos\theta)}.$$

2.5 Manifolds in R^m

It should be observed that our considerations can be extended to manifolds in R^n where

$$ds^2 = \sum_{ij} g_{ij} du_i du_j,$$

where $i, j = 1 \ldots n$.

In the same way that we defined surfaces in R^3 we can do the same in R^m where

$$\mathbf{x} = (x^1, x^2, \ldots, x^m), \quad d\mathbf{x} = (dx^1, dx^2, \ldots, dx^m).$$

Definition 2.12: A manifold of dimension n in R^m with $n \leq m$ is a mapping

$$\mathbf{x} : U \in R^n \rightarrow R^m, \quad n \leq m,$$

whose Jacobian at any point (with possible exception at a finite number of points of U) is of rank n.

$$\mathbf{x}(u^1, u^2, \ldots, u^n) = (x^1(u^1, u^2, \ldots, u^n), \ldots, x^m(u^1, u^2, \ldots, u^n)).$$

We then have

$$dx^i = \sum_{j=1}^{n} \frac{\partial x^i}{\partial u^j} du^j = \frac{\partial x^i}{\partial u^1} du^1 + \cdots + \frac{\partial x^i}{\partial u^n} du^n, \quad i = 1, \ldots m.$$

Hence,

$$dx = \left(\frac{\partial x^1}{\partial u^1} du^1 + \cdots + \frac{\partial x^1}{\partial u^n} du^n, \frac{\partial x^2}{\partial u^1} du^1 + \cdots \right.$$
$$\left. + \frac{\partial x^2}{\partial u^n} du^n, \ldots, \frac{\partial x^m}{\partial u^1} du^1 + \cdots + \frac{\partial x^m}{\partial u^n} du^n \right).$$

This can be rewritten as

$$dx = \left(\frac{\partial x^1}{\partial u^1}, \frac{\partial x^2}{\partial u^1}, \ldots, \frac{\partial x^m}{\partial u^1} \right) du^1 + \left(\frac{\partial x^1}{\partial u^2}, \frac{\partial x^2}{\partial u^2}, \ldots, \right.$$
$$\left. \frac{\partial x^m}{\partial u^2} \right) du^2 + \ldots + \left(\frac{\partial x^1}{\partial u^n}, \frac{\partial x^2}{\partial u^n}, \ldots, \frac{\partial x^m}{\partial u^n} \right) du^n.$$

Introducing

$$\mathbf{x}_{u^i} = \left(\frac{\partial x^1}{\partial u^i}, \frac{\partial x^2}{\partial u^i}, \ldots, \frac{\partial x^m}{\partial u^i} \right), \quad i = 1, \ldots, n.$$

we can write therefore

$$d\mathbf{x} = \mathbf{x}_{u^1} du^1 + \mathbf{x}_{u^2} du^2 + \cdots + \mathbf{x}_{u^n} du^n.$$

For the arc length, we therefore have

$$ds^2 = d\mathbf{x} \cdot d\mathbf{x} = (\mathbf{x}_{u^1} du^1 + \mathbf{x}_{u^2} du^2 + \cdots + \mathbf{x}_{u^n} du^n) \cdot (\mathbf{x}_{u^1} du^1 + \mathbf{x}_{u^2} du^2 + \cdots + \mathbf{x}_{u^n} du^n).$$

Hence

$$ds^2 = (\mathbf{x}_{u^1} \cdot \mathbf{x}_{u^1}) du^1 du^1 + (\mathbf{x}_{u^1} \cdot \mathbf{x}_{u^2}) du^1 du^2 + \cdots + (\mathbf{x}_{u^n} \cdot \mathbf{x}_{u^n}) du^n du^n,$$

or

$$ds^2 = \sum_{i,j=1}^{n} g_{ij} du^i du^j,$$

where

$$g_{ij} = \mathbf{x}_{u^i} \cdot \mathbf{x}_{u^j}.$$

Example 2.16: The sphere in R^4

A sphere in R^3 is defined as the set of all points whose distance from some point is constant. Similarly, we can define a sphere in R^n as the set of all points whose distance from some point in R^n is constant. Such a sphere in R^4 of radius R (a three dimensional surface) around $\mathbf{x} = 0$ can be parameterized using "spherical coordinates" as

$$\mathbf{x} = (R \sin \psi \sin \phi \cos \theta, R \sin \psi \sin \phi \sin \theta, R \sin \psi \cos \phi, R \cos \psi).$$

Hence

$$ds^2 = R^2 [d\psi^2 + \sin^2 \psi (d\phi^2 + \sin^2 \phi d\theta^2)].$$

The non-zero components of the metric tensor are

$$g_{11} = R^2, \quad g_{22} = R^2 \sin^2 \psi, \quad g_{33} = R^2 \sin^2 \psi \sin^2 \phi.$$

2.6   Tensors

To introduce the concept of tensors, we consider a (nonsingular) transformation $u^i = (\bar{x}^i) = u^i(x^1, \ldots, x^n)$ in R^n. Under this transformation, we have by the chain rule

$$du^i = d\bar{x}^i = \sum_j \frac{\partial u^i}{\partial x^j} dx^j.$$

On the other hand if we consider a vector field $A_i = \frac{\partial \psi}{\partial x^i}$ then under this coordinates transformation the components of A_i in will be

$$\bar{A}_i = \frac{\partial \psi}{\partial u_i} = \sum_j \frac{\partial \psi}{\partial x^j} \frac{\partial x^j}{\partial u_i} = \sum_j \frac{\partial x^j}{\partial u_i} A_j.$$

Thus, while the transformation coefficients for the differentials are $\left(\frac{\partial u^i}{\partial x^j}\right)$ the coefficients for the transformation of the components of A_i are given by the inverse transformation $\left(\frac{\partial x^j}{\partial u_i}\right)$.

Remark 2.3: Observe that

$$\frac{\partial u^i}{\partial u^j} = \sum_j \frac{\partial u^i}{\partial x^j} \frac{\partial x^j}{\partial u^i} = \delta^i_j.$$

where δ^i_j is the Kronecker delta which is equal zero if $i \neq j$ and 1 when $i = j$

For these reasons, we introduce now the following definitions

Definition 2.13: A set of n-quantities $(v^1, \ldots, v^n)$ in R^n is called a **contravariant vector** if under a transformation of the coordinates $\bar{x}^i = \bar{x}^i(x^1, \ldots, x^n)$ its components in the new coordinate system are given by

$$\bar{v}^i = \sum_j \frac{\partial \bar{x}^i}{\partial x^j} v^j.$$

Definition 2.14: A set of n-quantities $(w_1, \ldots, w_n)$ in R^n is called a **covariant vector** if under a transformation of the coordinates $\bar{x}^i = \bar{x}^i(x^1, \ldots, x^n)$ its components in the new coordinate system are given by

$$\bar{w}_i = \sum_j \frac{\partial x^j}{\partial \bar{x}^i} w_j.$$

Example 2.17: Consider the gradient of a function $f = f(x_1, \ldots, x_n)$.

$$\mathrm{grad}(f) = \nabla f = \left(\frac{\partial f}{\partial x_1}, \ldots, \frac{\partial f}{\partial x_n}\right).$$

If we make a coordinate transformation

$$x_i = x_i(u_1, \ldots, u_n),$$

then the expression for $\mathrm{grad}(f)$ in the new coordinate system is

$$\mathrm{grad}(f) = \nabla f = \left(\frac{\partial f}{\partial u_1}, \ldots, \frac{\partial f}{\partial u_n}\right).$$

By the chain rule we have

$$\frac{\partial f}{\partial u_i} = \sum_{j=1}^{n} \frac{\partial f}{\partial x_j} \frac{\partial x_j}{\partial u_i}.$$

Thus, if we let $T_j = \frac{\partial f}{\partial x_j}$ and $\bar{T}_i = \frac{\partial f}{\partial u_i}$, we can rewrite the chain rule for the transformation as

$$\bar{T}_i = \sum_{j=1}^{n} T_j \frac{\partial x_j}{\partial u_i}.$$

This demonstrates clearly that the gradient of a function is a covariant tensor of rank one.

Example 2.18: We showed that

$$d\mathbf{x} = (dx^1, \ldots, dx^n),$$

is a contravariant vector. Therefore, $T^{ij} = dx^i dx^j$ is a contravariant tensor of rank 2. However,

$$ds^2 = g_{ij} dx^i dx^j,$$

is obviously a scalar. Therefore, it follows from the quotient theorem that g_{ij} is a covariant tensor of rank 2.

2.6.1   Conventions

1. The components of a contravariant vector will be denoted by a **superscript**. The components of a covariant vector will be denoted by a **subscript**.

2. (Einstein summation convention) When the same index appear in a formula as a superscript and a subscript, a summation over that index is assumed.

3. An index that appears in a formula as $\frac{\partial}{\partial x^i}$ will be considered as a **subscript** (even though it is written as a superscript).

4. A scalar (tensor of rank zero) is a quantity whose value at a (fixed) point in space does not change under a coordinate transformation.

Example 2.19: From now on we shall write $\frac{\partial x^j}{\partial \bar{x}^i} w_j$ for

$$\sum_j \frac{\partial x^j}{\partial \bar{x}^i} w_j.$$

2.6.2 Tensors in General

Now that we have defined contravariant and covariant vectors, and we can compound these definitions and define tensors of any rank.

Example 2.20: A second rank contraviariant tensor T^{ij} will transform as

$$\bar{T}^{mn} = \frac{\partial \bar{x}^m}{\partial x^i} \frac{\partial \bar{x}^n}{\partial x^j} T^{ij}.$$

Similarly, a second rank covariant tensor will transform as

$$\bar{T}_{mn} = \frac{\partial x^i}{\partial \bar{x}^m} \frac{\partial x^j}{\partial \bar{x}^n} T_{ij}.$$

A third rank mixed tensor 2-covariant and 1-contravariant will transform as

$$\bar{T}^k_{mn} = \frac{\partial x^i}{\partial \bar{x}^m} \frac{\partial x^j}{\partial \bar{x}^n} \frac{\partial \bar{x}^k}{\partial x^s} T^s_{ij}.$$

Based on these transformation rules and the results from the previous section, it is clear that g_{ij} is a covariant second rank tensor.

We can define now several operations on tensors that preserve their transformation properties.

1. Two tensors of the same rank and type can be multiplied by a scalar and added to obtain a tensor of the same rank and type. For example, $S^i_j = \alpha T^i_j + \beta R^i_j$ is a mixed tensor of rank 2.

2. Two tensors of rank (m_1, n_1),(m_2, n_2) can be multiplied to obtain a tensor of rank $(m_1 + m_2, n_1 + n_2)$.

3. (Contraction) Let T be of rank (m_1, n_1) and R of rank (m_2, n_2). If we multiply these tensors and sum over one contravariant index and one covariant index, we obtain a tensor S of rank $(m_1 + m_2 - 1, n_1 + n_2 - 1)$. For example $S^j = T^{ij} R_i$ is a contravariant tensor of rank one. This theorem can be used also in "reverse". If we know that S and R are tensors, then it follows that T is also a tensor (with the proper ranks). This theorem is known also as "the quotient theorem for tensors".

Example 2.21: The metric tensor g_{ij} is a covariant tensor of rank two. On the other hand, its "inverse" g^{ij} is a contravariant tensor of rank two. These two tensors are used "routinely" to "raise and lower" the indices of other tensors. For example, if v_i is a covariant vector, then $w^k = g^{ki} v_i$ is a contravariant vector.

Example 2.22: To show that g^{ij} is a contravariant tensor of rank two, we consider a contavariant vector v^i. Since g_{ij} is a covariant tensor of rank two, it follows that $w_k = g_{ij}v^i$ is a covariant vector. Written explicitly we have

$$\begin{pmatrix} w_1 \\ w_2 \end{pmatrix} = \begin{pmatrix} g_{11} & g_{12} \\ g_{21} & g_{22} \end{pmatrix} \begin{pmatrix} v^1 \\ v^2 \end{pmatrix}. \tag{2.4}$$

Solving this equation for (v^1, v^2) we have

$$\begin{pmatrix} v^1 \\ v^2 \end{pmatrix} = \frac{1}{g} \begin{pmatrix} g_{22} & -g_{12} \\ -g_{21} & g_{11} \end{pmatrix} \begin{pmatrix} w_1 \\ w_2 \end{pmatrix}, \tag{2.5}$$

where g is the determinant of the matrix g_{ij}. Hence, by the quotient theorem, it follows that

$$g^{11} = \frac{1}{g} g_{22}, \quad g^{22} = \frac{1}{g} g_{11}, \quad g^{12} = g^{21} = -\frac{1}{g} g_{12},$$

is a contravariant tensor of rank two.

The metric tensor g_{ij} is a covariant tensor of rank two. On the other hand, its "inverse" g^{ij} is a contravariant tensor of rank two. These two tensors are used to "raise and lower" the indices of other tensors. For example if v_i is a covariant vector then $w^k = g^{ki}T_i$ is a contravariant vector.

Remark 2.4: Tensors are important from a physical point of view since equations expressing a physical law in terms of equality between two tensors are independent of the coordinate systems (observers) used to express them.

2.6.3 Relative Tensors

A relative tensor transforms under a change of coordinates as a regular tensor but in addition a "weight" is added as a coefficient.

For example, if T^α_β is a mixed second rank tensor then its transformation under a change of coordinates is given by

$$\bar{T}^\alpha_\beta = \frac{\partial \bar{x}^\alpha}{\partial x^a} \frac{\partial x^b}{\partial \bar{x}^\beta} \det(J)^m T^a_b,$$

where m is an integer (positive or negative) and J is the Jacobian of the transformation.

$$J = \left(\frac{\partial \bar{x}^i}{\partial x^j} \right).$$

We observe also that J^{-1} is given by

$$J^{-1} = \left(\frac{\partial x^j}{\partial \bar{x}^i} \right),$$

since

$$J J^{-1} = \left(\frac{\partial \bar{x}^i}{\partial \bar{x}^j} \right) = (\delta^i_j).$$

When the "weight" $m = -1$ such relative tensors are referred to as *densities*.

How one can "generate" a physically meaningful relative tensor on a Riemannian manifold?

To this end, we consider the metric tensor g_{ij} which transforms as

$$\bar{g}_{ij} = \frac{\partial x^a}{\partial \bar{x}^i} \frac{\partial x^b}{\partial \bar{x}^j} g_{ab}.$$

This expression can re-interpreted in terms of matrix multiplication, and hence when we take the determinant of these matrices, we obtain

$$\bar{g} = (\det(J^{-1}))^2 g,$$

where g and $\bar{g}$ are the determinants of the metric tensors before and after the coordinate transformation. Observe that **that this result holds for orientable and non-orientable manifolds and the sign of g remains invariant under coordinate transformations**

Assuming both g and $\bar{g}$ are positive, we can take the square root on both sides of this equation to obtain

$$\sqrt{\bar{g}} = \sqrt{g} |\det(J)^{-1}|. \tag{2.6}$$

If the manifold is orientable $\det(J)^{-1} > 0$ and $\sqrt{\bar{g}}$ is a scalar (tensor) density, i.e., we can get rid of the absolute sign on $\det(J)^{-1}$ in (2.6).

As a side remark, we note that if the determinant of g_{ij} is negative (as in Minkowiski space or General Relativity), we replace $\sqrt{g}$ by $\sqrt{-g}$ in the previous formulas.

Let $\mathbf{T}$ be a tensor (on an orientable manifold) then it now clear that $\mathbf{S} = \mathbf{T}\sqrt{g}$ is a density. For example, if $\mathbf{T} = T_{ij}$ then

$$\bar{S}_{ij} = \bar{T}_{ij}\sqrt{\bar{g}} = \frac{\partial x^a}{\partial \bar{x}^i} \frac{\partial x^b}{\partial \bar{x}^j} T_{ab}\sqrt{g} \det(J)^{-1} = \det(J)^{-1} \frac{\partial x^a}{\partial \bar{x}^i} \frac{\partial x^b}{\partial \bar{x}^j} S_{ab}.$$

Appendix 2A: Maple and MATLAB® Programs

In this chapter, we provide a maple program for the torus that can be used to compute the metric tensor and the second and third Fundamental forms. This program can be adapted to other surfaces by just changing the second line of the program to the equation of the desired surface in R^3 the name of the program is "torus.mw".

A second program "Mobius.mw" calculates the expressions for the tangent vectors to the Mobius strip and its Normal vector N. It then demonstrates that this surface is not orientable by showing that $\mathbf{N}(0,0) \neq \mathbf{N}(2\pi,0)$

Two MATLAB® programs "torus.m" and "hyper.m" are available also. These are used to plot the torus and one sheet hyperboloid and can be modified easily to plot other surfaces in R^3.

Exercises

1. Plot some of the surfaces enumerated in the first section of this chapter.

2. Adopt the maple program "torus.mw' to calculate the metric tensor, the second and third fundamental forms for the surfaces enumerated in the first section of this chapter.

3. Calculate the contravariant metric tensor for each of the surfaces in Exercise 2.

4. A surface is given in the form

$$\mathbf{x} = (x, y, f(x, y)),$$

where $f(x, y)$ is a smooth function. Calculate the metric tensor and the second and third fundamental forms of this surface.

5. A surface of revolution is is created by revolving the curve $z = f(x)$ in the x-z plane around the z axis. The resulting surface is

$$\mathbf{x}(r, \theta) = (r \cos \theta, r \sin \theta, f(r)),$$

where θ is the rotation angle around the z-axis. Evaluate the metric tensor of this surface.

6. Compute the metric tensor for the three-dimensional ellipsoid

$$\mathbf{x} = (a \sin \psi \sin \phi \cos \theta, b \sin \psi \sin \phi \sin \theta, c \sin \psi \cos \phi, d \cos \psi) = 1.$$

7. Compute the metric tensor for the three-dimensional hyperboloid of one sheet

$$\left(\frac{x}{a}\right)^2 + \left(\frac{y}{b}\right)^2 + \left(\frac{z}{c}\right)^2 - \left(\frac{w}{d}\right)^2 = 1.$$

Another representation of this surface is

$$\mathbf{x} = (a\cosh(u)\sin(\phi)\cos(\theta), b\cosh(u)\sin(\phi)\sin(\theta),$$
$$c*\cosh(u)\cos(\phi), d\sinh(u)).$$

Verify this representation and compute the corresponding metric tensor.

8. Compute the second fundamental form of the two-dimensional sphere and its Gaussian curvature.

9. Let T_i be a covariant tensor. Prove that $g^{ji}T_i$ is a conravariant tensor.

10. A two-dimensional surface in three dimensions is defined by $z = f(x,y)$. Derive the expressions for the Gaussian and mean curvature of this surface.

3

Tensor Analysis on Riemann Manifolds

3.1 Geodesics

In the plane, the shortest path between two points is a straight line. Geodesics are curves that are the "equivalent" to straight lines on Riemann manifold. That is they provide the "shortest" path between two points on the manifold.

On a Riemann manifold, we have

$$ds^2 = g_{ij}du^i du^j.$$

Hence for a curve $u^i = u^i(t)$ on the manifold, we have

$$\left(\frac{ds}{dt}\right)^2 = g_{ij}(\mathbf{u})\frac{du^i}{dt}\frac{du^j}{dt}.$$

The distance between two points $P_0 = (u^1(t_0), \ldots, u^n(t_0))$ and $P_1 = (u^1(t_1), \ldots, u^n(t_1))$ on this curve is

$$s = \int_{t_0}^{t_1} \sqrt{\left[g_{ij}(\mathbf{u})\frac{du^i}{dt}\frac{du^j}{dt}\right]}\, dt.$$

Suppose now that we found the geodesic curve $(u^1(t), \ldots, u^n(t))$ connecting P_0 and P_1. Any other curve joining these two points and close to the geodesics curve can represented as

$$\bar{u}^i(t) = u^i(t) + \epsilon w^i(t), \qquad (3.1)$$

where $w(t_0) = w(t_1) = 0$ and $\epsilon \ll 1$. The distance between the points P_0, P_1 along this curve is

$$\bar{s} = \int_{t_0}^{t_1} \sqrt{\left[g_{ij}(\bar{\mathbf{u}})\frac{d\bar{u}^i}{dt}\frac{d\bar{u}^j}{dt}\right]}\, dt. \qquad (3.2)$$

Substituting (3.1) in (3.2) and keeping only first powers of ϵ, we obtain the following approximation for $\bar{s} - s$

$$\bar{s} - s = \frac{\epsilon}{2}\int_{s_0}^{s_1} \left\{\frac{\partial g_{ij}}{\partial u^k}\frac{du^i}{ds}\frac{du^j}{ds} - 2\frac{d}{ds}\left(g_{kj}\frac{du^j}{ds}\right)\right\} w^k ds, \qquad (3.3)$$

DOI: 10.1201/9781003587422-3

where ds represents the (infinitesimal) arc length along the geodesic. Since w^k are arbitrary, the condition for extremum is that the integrand in (3.3) is zero.

$$\frac{d}{ds}\left(g_{kj}\frac{du^j}{ds}\right) - \frac{1}{2}\frac{\partial g_{ij}}{\partial u^k}\frac{du^i}{ds}\frac{du^j}{ds} = 0, \qquad (3.4)$$

for $k = 1, \ldots, n$. Switching the roles of the indices i, k in this expression (that is making the substitutions $i \to k$ and $k \to i$) we obtain

$$\frac{d}{ds}\left(g_{ij}\frac{du^j}{ds}\right) - \frac{1}{2}\frac{\partial g_{jk}}{\partial u^i}\frac{du^k}{ds}\frac{du^j}{ds} = 0. \qquad (3.5)$$

Hence

$$g_{ij}\frac{d^2 u^j}{ds^2} + \frac{\partial g_{ij}}{\partial u^k}\frac{du^k}{ds}\frac{du^j}{ds} - \frac{1}{2}\frac{\partial g_{jk}}{\partial u^i}\frac{du^k}{ds}\frac{du^j}{ds} = 0. \qquad (3.6)$$

We symmetrize this result using the fact that the summation indices are dummy, and we can switch the roles of the indices k and j (which are being summed over)

$$\frac{\partial g_{ij}}{\partial u^k}\frac{du^j}{ds}\frac{du^k}{ds} = \frac{\partial g_{ik}}{\partial u^j}\frac{du^j}{ds}\frac{du^k}{ds}. \qquad (3.7)$$

We finally obtain

$$g_{ik}\frac{d^2 u^k}{ds^2} + \frac{1}{2}\left(\frac{\partial g_{ij}}{\partial u^k} + \frac{\partial g_{ik}}{\partial u^j} - \frac{\partial g_{jk}}{\partial u^i}\right)\frac{du^j}{ds}\frac{du^k}{ds} = 0. \qquad (3.8)$$

Motivated by this expression, we now introduce the Christoffel symbols of the first and second kind

$$\Gamma_{jki} = (jk, i) = \frac{1}{2}\left(\frac{\partial g_{ij}}{\partial u^k} + \frac{\partial g_{ik}}{\partial u^j} - \frac{\partial g_{jk}}{\partial u^i}\right). \qquad (3.9)$$

$$\Gamma_{jk}^m = \begin{pmatrix} m \\ jk \end{pmatrix} = \frac{1}{2}g^{mi}\left(\frac{\partial g_{ij}}{\partial u^k} + \frac{\partial g_{ik}}{\partial u^j} - \frac{\partial g_{jk}}{\partial u^i}\right) = g^{mi}\Gamma_{jki}. \qquad (3.10)$$

Using this notation, the equations of a geodesic can written in one of the following forms:

$$g_{ik}\frac{d^2 u^k}{ds^2} + \Gamma_{jki}\frac{du^j}{ds}\frac{du^k}{ds} = 0, \qquad (3.11)$$

$$\frac{d^2 u^m}{ds^2} + \Gamma_{jk}^m\frac{du^j}{ds}\frac{du^k}{ds} = 0. \qquad (3.12)$$

The second equation can be obtained from the first equation by multiplying the first by g^{im} and summing over i.

To derive the transformation properties of the Christoffel symbols under a change of the coordinate system, we observe that in the derivation of (3.12) no special assumptions were made about the coordinate system. Hence, if a different coordinate system $\bar{u}^i$ is used, we shall obtain

$$\frac{d^2\bar{u}^m}{ds^2} + \bar{\Gamma}^m_{jk}\frac{d\bar{u}^j}{ds}\frac{d\bar{u}^k}{ds} = 0. \tag{3.13}$$

where $\bar{\Gamma}^m_{jk}$ is the expression of the Christoffel symbols in the new coordinate system. Using the chain-rule, we obtain after some algebra that

$$\Gamma_{\alpha\beta\rho} = \left[\bar{\Gamma}_{\lambda\nu\tau}\frac{\partial\bar{u}^\lambda}{\partial u^\alpha}\frac{\partial\bar{u}^\nu}{\partial u^\beta} + \bar{g}_{\sigma\tau}\frac{\partial^2\bar{u}^\sigma}{\partial u^\alpha\partial u^\beta}\right]\frac{\partial\bar{u}^\tau}{\partial u^\rho}. \tag{3.14}$$

It follows from this formula that the Christoffel symbols are not tensorial quantities. The second term will vanish; however, if we restrict ourselves to linear transformations. Under this restriction, the Christoffel symbols will become tensorial quantities.

3.2   Examples of Geodesics

Example 3.1: Show that straight lines are the geodesics in R^n

   Solution: In R^n we have (by Pythagoras Theorem)

$$ds^2 = (dx_1)^2 + \cdots + (dx_n)^2,$$

i.e. $g_{ij} = \delta_{ij}$ therefore (3.4) implies that

$$\frac{d^2x_i}{ds^2} = 0, \quad i = 1,\ldots,n.$$

Hence

$$x_i(s) = A_i s + B_i,$$

which is an equation of a (straight)line.

   Example 3.2: Find the geodesics on the cylinder

$$\mathbf{x}(\theta, u) = (r\cos\theta, r\sin\theta, u).$$

(The Cartesian equation of this surface is $x^2 + y^2 = r^2$).

   Solution: The tangent vectors to the surface are

$$\mathbf{x}_\theta = (-r\sin\theta, r\cos\theta, 0), \quad \mathbf{x}_u = (0, 0, 1).$$

Hence $g_{11} = r^2$, $g_{12} = 0$, $g_{22} = 1$. The differential equations for the geodesics are

$$\frac{d}{ds}\left(r^2\frac{d\theta}{ds}\right) = 0, \quad \frac{d}{ds}\left(\frac{du}{ds}\right) = 0.$$

The equation of a geodesic is given by

$$\theta = As + B, \quad u = Cs + D,$$

where A, B, C, D are constants. The curves on the cylinder are

$$\mathbf{x}(s) = (r\cos(As + B), r\sin(As + b), Cs + D).$$

Example 3.3: Show that the great circles on S^2 (sphere in R^3) are geodesics on this surface.

Solution: The representation of S^2 in spherical coordinates is

$$\mathbf{x}(\theta, \phi) = (r\sin\phi\cos\theta, r\sin\phi\sin\theta, r\cos\phi).$$

The tangent vectors are

$$\mathbf{x}_\theta = (-r\sin\phi\sin\theta, r\sin\phi\cos\theta, 0), \quad \mathbf{x}_\phi = (r\cos\phi\cos\theta, r\cos\phi\sin\theta, -r\sin\phi).$$

Therefore,

$$g_{11} = \mathbf{x}_\theta \cdot \mathbf{x}_\theta = r^2\sin^2\phi, \quad g_{12} = \mathbf{x}_\theta \cdot \mathbf{x}_\phi = 0, \quad g_{22} = \mathbf{x}_\phi \cdot \mathbf{x}_\phi = r^2.$$

The differential equations for the geodesics (from (3.4) are

$$\frac{d}{ds}\left(r^2\sin^2\phi\frac{d\theta}{ds}\right) = 0, \quad k = 1,$$

$$\frac{d}{ds}\left(r^2\frac{d\phi}{ds}\right) - r^2\sin\phi\cos\phi\left(\frac{d\theta}{ds}\right)^2 = 0, \quad k = 2,$$

Particular solutions of this system of differential equations are $\theta = $ *constant*, $r\phi = s$ or $\phi = \frac{\pi}{2}$, $r\theta = s$. These solutions represent respectively the "meridians" and the "equator" with respect to the point chosen as a pole. Since the choice of the pole is arbitrary, this shows that any great circle is a geodesic on S^2. Observe that the "small circles" $\phi = $ *constant*, $\phi \neq \frac{\pi}{2}$ are not geodesics.

Example 3.4: Compute the equation of the geodesics on the hyperboloid of one sheet

$$\mathbf{x}(\phi, \theta) = (\cosh\phi\cos\theta, \cosh\phi\sin\theta, \sinh\phi).$$

Solution: The tangent vectors on this surface are

$$\mathbf{x}_\phi = (\sinh\phi\cos\theta, \sinh\phi\sin\theta, \cosh\phi), \quad \mathbf{x}_\theta = (-(\cosh\phi\sin\theta, \cosh\phi\cos\theta, 0).$$

Hence

$$g_{11} = \cosh(2\phi), \quad g_{12} = g_{21} = 0, \quad g_{22} = \cosh^2\phi.$$

The differential equation for the geodesics (3.4) yields then

$$\frac{d}{ds}(\cosh(2\phi)\frac{d\phi}{ds}) - \frac{1}{2}\frac{d}{d\phi}(\cosh(2\phi))(\frac{d\phi}{ds})^2 - \frac{1}{2}\frac{d}{d\phi}(\cosh^2\phi)(\frac{d\theta}{ds})^2 = 0, \quad k = 1,$$
(3.15)

$$\frac{d}{ds}(\cosh^2\phi\frac{d\theta}{ds}) = 0.$$
(3.16)

Particular solutions of this system can be obtained in the following cases:

A. $\phi = constant$. In this case, Eq. (3.15) is satisfied by default and (3.16) reduces to

$$\frac{d^2\theta}{ds^2} = 0,$$

whose solution is

$$\theta(s) = As + B.$$

B. $\theta = constant$. Eq. (3.16) is satisfied and (3.15) reduces to

$$\frac{d^2\phi}{ds^2} + \tanh(2\phi)(\frac{d\phi}{ds})^2 = 0.$$
(3.17)

A solution of this equation is given by

$$\phi(s) = \operatorname{arcsinh}(C_1 s + C_2).$$

A more general solution of (3.17) can be obtained in terms of elliptic functions.

Example 3.5: Compute the equation of the geodesics on the Cone,

$$\mathbf{x}(\theta, u) = (u\cos\theta, u\sin\theta, u).$$

(In Cartesian coordinates the equation off this cone is $x^2 + y^2 = z^2$).

The tangent vectors are

$$\mathbf{x}_\theta = (-u\sin\theta, u\cos\theta, 0), \quad \mathbf{x}_u = (\cos\theta, \sin\theta, 1).$$

Hence

$$g_{11} = u^2, \quad g_{12} = g_{21} = 0, \quad g_{22} = 2.$$

Using (3.4) we obtain the following equations for the geodesics

$$\frac{d}{ds}(u^2\frac{d\theta}{ds}) = 0, \quad k = 1,$$

$$2\frac{d^2u}{ds^2} - u(\frac{d\theta}{ds})^2 = 0, \quad k = 2.$$

From the first equation, we obtain

$$\frac{d\theta}{ds} = \frac{c}{u^2},$$
(3.18)

where c is a constant. Substituting this result in the second equation leads to

$$2\frac{d^2u}{ds^2} - \frac{c^2}{u^3} = 0.$$

Multiplying this equation by u' we have

$$2u''u' = \frac{c^2u'}{u^3},$$

(where primes denote differentiation with respect to s). Integrating this equation with respect to s (using the fact that $2u''u' = ((u')^2)')$

$$u' = \frac{\sqrt{2du^2 - c^2}}{\sqrt{2}u}.$$

Finally, by integrating this equation, we obtain

$$s = \frac{1}{\sqrt{2}}\frac{\sqrt{2du^2 - c^2}}{d} + e,$$

where e is an integration constant. To compute $\theta(s)$, we substitute this result in (3.18) and integrate. This yields

$$\theta = \sqrt{2}\arctan\left[\frac{\sqrt{2d}}{c}(s - c)\right] + C_1.$$

3.3 Covariant Differentiation

To see the need to redefine the differentiation operation on manifolds, we consider a contravariant vector T^i. Under a change of the coordinate system, the new components of this vector will be

$$\bar{T}^j = T^i\frac{\partial\bar{u}^j}{\partial u^i}.$$

If we differentiate $\bar{T}^j$ we obtain

$$\frac{\partial\bar{T}^\alpha}{\partial\bar{u}^\rho} = \frac{\partial T^\beta}{\partial u^\sigma}\frac{\partial u^\sigma}{\partial\bar{u}^\rho}\frac{\partial\bar{u}^\alpha}{\partial u^\beta} + T^\kappa\frac{\partial^2\bar{u}^\alpha}{\partial u^\kappa\partial u^\sigma}\frac{\partial u^\sigma}{\partial\bar{u}^\rho}. \tag{3.19}$$

It follows then that in general the regular differentiation operator does not "conserve" the tensorial character of the quantities it operates on. We must therefore redefine the differentiation operation so that tensors remain tensors under this operation. This "new" definition of the differentiation operation will be referred to as "covariant differentiation".

To see how this can be done, we use (3.14) to obtain

$$\frac{\partial^2 \bar{u}^\alpha}{\partial u^\kappa \partial u^\sigma} = \bar{\Gamma}^\beta_{\kappa\sigma} \frac{\partial \bar{u}^\alpha}{\partial u^\beta} - \Gamma^\alpha_{\tau\nu} \frac{\partial \bar{u}^\tau}{\partial u^\kappa} \frac{\partial \bar{u}^\nu}{\partial u^\alpha}. \tag{3.20}$$

Substituting this in (3.3), we obtain after some algebra

$$\frac{\partial \bar{T}^\alpha}{\partial \bar{u}^\rho} + \bar{\Gamma}^\alpha_{\tau\rho} \bar{T}^\tau = \left(\frac{\partial T^\beta}{\partial \bar{u}^\sigma} + \Gamma^\beta_{\kappa\sigma} T^\kappa \right) \frac{\partial u^\sigma}{\partial \bar{u}^\rho} \frac{\partial \bar{u}^\alpha}{\partial u^\beta}. \tag{3.21}$$

Hence, if we introduce

$$\bar{T}^\alpha_{\|\rho} = \frac{\partial \bar{T}^\alpha}{\partial \bar{u}^\rho} + \bar{T}^\tau \bar{\Gamma}^\alpha_{\tau\rho},$$

and

$$T^\beta_{\|\sigma} = \frac{\partial T^\beta}{\partial u^\sigma} + T^\kappa \bar{\Gamma}^\beta_{\kappa\sigma},$$

then we have

$$\bar{T}^\alpha_{\|\rho} = T^\beta_{\|\sigma} \frac{\partial u^\sigma}{\partial \bar{u}^\rho} \frac{\partial \bar{u}^\alpha}{\partial u^\beta},$$

that is $T^\alpha_{\|\rho}$ is a second rank mixed tensor (with respect to the metric tensor $g_{\alpha\beta}$). This is the definition of the covariant derivative of a first rank tensor, which will be symbolized by a parallel line. From now on, the regular derivative will be symbolized by one line.

A geometrical method to derive this formula (and similar for higher rank tensors) is to observe that geodesics are intrinsic to the manifold and and are independent of the coordinate system used. Let tangent vector to a geodesic be $\lambda^i = du^i/ds$. Contracting this vector with T_i (to obtain a scalar) and using (3.9) and (3.10), we have

$$\begin{aligned} \frac{d(T_\nu \lambda^\nu)}{ds} &= \frac{d(g_{\mu\nu} T^\mu \lambda^\nu)}{ds} \\ &= \left\{ T^\mu \frac{\partial g_{\mu\nu}}{\partial u^\sigma} + g_{\mu\nu} \frac{\partial T^\mu}{\partial u^\sigma} - \Gamma^\tau_{\sigma\nu} g_{\mu\tau} T^\mu \right\} \lambda^\nu \lambda^\sigma \\ &= \left\{ g_{\mu\nu} \frac{\partial T^\mu}{\partial u^\sigma} + \left[\frac{\partial g_{\mu\nu}}{\partial u^\sigma} - \Gamma_{\sigma\nu\mu} \right] T^\mu \right\} \lambda^\nu \lambda^\sigma \\ &= \left[\frac{\partial T^\mu}{\partial u^\sigma} + \Gamma^\mu_{\sigma\nu} T^\nu \right] \lambda_\mu \lambda^\sigma. \end{aligned} \tag{3.22}$$

This demonstrates that

$$\bar{T}^\mu_{\|\sigma} = \frac{\partial T^\mu}{\partial u^\sigma} + \Gamma^\mu_{\sigma\nu} T^\nu,$$

is a mixed second rank tensor.

In a similar manner, we can derive the following table for the covariant derivatives;

$$T_{\mu\|\sigma} = \frac{\partial T_\mu}{\partial u^\sigma} - \Gamma^\nu_{\sigma\mu} T_\nu,$$

$$T_{\mu\nu\|\sigma} = \frac{\partial T_{\mu\nu}}{\partial u^\sigma} - \Gamma^\tau_{\mu\sigma} T_{\tau\nu} - \Gamma^\tau_{\nu\sigma} T_{\mu\tau},$$

$$T^\mu_{\|\nu} = \frac{\partial T^\mu}{\partial u^\nu} + \Gamma^\mu_{\nu\sigma} T^\sigma,$$

$$T^{\lambda\nu}_{\|\mu} = \frac{\partial T^{\lambda\nu}}{\partial u^\mu} + \Gamma^\lambda_{\mu\sigma} T^{\sigma\nu} + \Gamma^\nu_{\mu\sigma} T^{\lambda\sigma},$$

$$T^\lambda_{\nu\|\mu} = \frac{\partial T^\lambda_\nu}{\partial u^\mu} + \Gamma^\lambda_{\mu\sigma} T^\sigma_\nu - \Gamma^\sigma_{\mu\nu} T^\lambda_\sigma.$$

Example 3.6: Show that the covariant derivative of the Kronecker δ is zero viz.

$$\delta^\mu_{\nu\|\lambda} = 0.$$

Solution: To show this, we use the formula above for the covariant derivative of the mixed second rank tensor with $\lambda \to \mu$, $\mu \to \lambda$.

$$\delta^\mu_{\nu\|\lambda} = \frac{\partial \delta^\mu_\nu}{\partial u^\lambda} + \Gamma^\mu_{\sigma\lambda} \delta^\sigma_\nu - \Gamma^\sigma_{\lambda\nu} \delta^\mu_\sigma = \Gamma^\mu_{\nu\lambda} - \Gamma^\mu_{\lambda\nu} = 0.$$

Remark 3.1: $g_{\mu\nu\|\lambda} = 0$.

In fact

$$g_{\mu\nu\|\lambda} = \frac{\partial g_{\mu\nu}}{\partial u^\lambda} - \Gamma^\sigma_{\mu\lambda} g_{\sigma\nu} - \Gamma^\sigma_{\nu\lambda} g_{\mu\sigma} \tag{3.23}$$

$$= \frac{\partial g_{\mu\nu}}{\partial u^\lambda} - \Gamma_{\mu\lambda\nu} - \Gamma_{\nu\lambda\mu}$$

$$= \frac{\partial g_{\mu\nu}}{\partial u^\lambda} - \frac{1}{2}\left[\frac{\partial g_{\mu\nu}}{\partial u^\lambda} + \frac{\partial g_{\lambda\nu}}{\partial u^\mu} - \frac{\partial g_{\mu\lambda}}{\partial u^\nu}\right] - \frac{1}{2}\left[\frac{\partial g_{\nu\mu}}{\partial u^\lambda} + \frac{\partial g_{\lambda\mu}}{\partial u^\nu} - \frac{\partial g_{\nu\lambda}}{\partial u^\mu}\right] = 0.$$

Remark 3.2: The divergence operator of a tensor is defined by covariant differentiation and contraction (of one of the indices) with the differentiation index. For example,

$$T^\lambda_{\|\lambda} = \frac{\partial T^\lambda}{\partial u^\lambda} + \Gamma^\lambda_{\nu\lambda} T^\nu = \frac{1}{\sqrt{g}} \frac{\partial(\sqrt{g} T^\lambda)}{\partial u^\lambda},$$

where g is the determinant of the metric tensor. Similarly,

$$T^{\mu\lambda}_{\|\lambda} = \frac{\partial T^{\mu\lambda}}{\partial u^\lambda} + \Gamma^\lambda_{\sigma\lambda} T^{\mu\sigma} + \Gamma^\mu_{\sigma\lambda} T^{\sigma\lambda} = \frac{1}{\sqrt{g}} \frac{\partial(\sqrt{g} T^{\mu\lambda})}{\partial u^\lambda} + \Gamma^\mu_{\sigma\lambda} T^{\sigma\lambda}.$$

3.4 Riemann and Ricci Tensors

While considering second-order covariant derivatives of a vector field, Riemann found a tensor of degree four, which depends only on the metric tensor. This tensor is

$$R^{\sigma}_{\lambda\nu\mu} = \frac{\partial}{\partial u^{\mu}}\Gamma^{\sigma}_{\lambda\nu} - \frac{\partial}{\partial u^{\nu}}\Gamma^{\sigma}_{\lambda\mu} + \Gamma^{\tau}_{\lambda\nu}\Gamma^{\sigma}_{\tau\mu} - \Gamma^{\tau}_{\lambda\mu}\Gamma^{\sigma}_{\tau\nu}. \qquad (3.24)$$

Since this tensor depends only on the metric tensor, it is intrinsic to the geometrical properties of the manifold. In fact, it is referred to as "the curvature tensor" because it generalizes this intuitive concept for surfaces in three dimensions.

Riemann (or Riemann-Christoffel) tensor has some indicial symmetries e.g

$$R_{\kappa\lambda\mu\nu} = -R_{\lambda\kappa\mu\nu}, \quad R_{\kappa\lambda\mu\nu} = R_{\kappa\lambda\nu\mu}, \quad R_{\kappa\lambda\mu\nu} = R_{\mu\nu\kappa\lambda}.$$

From a physical point of view, the most important tensor is the Ricci tensor, which is obtained from the Riemann tensor by contraction

$$R_{\lambda\mu} = R^{\nu}_{\lambda\mu\nu}.$$

Einstein theory of general relativity relates this tensor to the Energy-Momentum tensor. However, this tensor found many mathematical applications in the theory of partial differential equations and smoothing of differential manifold (what is called "Ricci flow").

Out of the Ricci tensor, we obtain by contraction

$$R = R^{\nu}_{\nu} = g^{\nu\sigma}R_{\sigma\nu},$$

which is called the "curvature invariant" (or "Ricci scalar") on the manifold.

Using the Ricci tensor and R, we can define the Einstein tensor

$$E^{\sigma}_{\nu} = R^{\sigma}_{\nu} - \frac{1}{2}\delta^{\sigma}_{\nu}(R - 2\Lambda),$$

where Λ is a constant. The (covariant) divergence of this tensor is zero

$$E^{\sigma}_{\mu\|\sigma} = 0.$$

Example 3.7: The metric on a two-dimensional sphere of radius R using spherical coordinates is given by

$$ds^2 = r^2[(d\theta)^2 + \sin^2\theta(d\phi)^2],$$

i.e. $g_{11} = r^2$, $g_{12} = 0$, $g_{22} = r^2 \sin^2 \theta$, $g = r^4 \sin^2 \theta$. The nonzero Christoffel symbols are

$$\Gamma_{12,2} = \Gamma_{21,2} = -\Gamma_{22,1} = r^2 \sin \theta \cos \theta,$$

$$\Gamma_{22}^1 = -\frac{1}{2} \sin 2\theta, \quad \Gamma_{12}^2 = \Gamma_{21}^2 = \cot \theta.$$

The Riemann tensor has only one nonzero component

$$R_{1221} = -r^2 \sin^2 \theta.$$

The Ricci tensor has two-nonzero components

$$R_{11} = -1, \quad R_{22} = -\sin^2 \theta,$$

and the invariant curvature R_ν^ν is

$$R_\nu^\nu = g^{11} R_{11} + g^{22} R_{22} = -2/r^2,$$

which corresponds intuitively to the notion of curvature of the sphere.

Example 3.8: For spaces with metric

$$ds^2 = \sum \epsilon_\nu (dx^\nu)^2,$$

where $\epsilon = \pm 1$ all the Christoffel symbols are zero and consequently the Riemann tensor etc are all zero. Such spaces are called flat.

To provide a further geometrical meaning to the Riemann tensor, we consider the second-order covariant derivatives of a covariant vector.

In R^n the mixed derivatives of such a vector are independent (subject to proper smoothness assumptions) of the differentiation order, e.g.,

$$\frac{\partial^2 v_i}{\partial x \partial y} = \frac{\partial^2 v_i}{\partial y \partial x}.$$

On a general manifold, we have

$$v_{i\|j} = \frac{\partial v_i}{\partial u^j} - v_m \Gamma_{ij}^m,$$

and hence

$$v_{i\|jk} = \frac{\partial v_i}{\partial u^j \partial u^k} - v_{m\|j} \Gamma_{ik}^m - v_{i\|m} \Gamma_{jk}^m.$$

Rewriting this explicitly, we obtain after some algebra that

$$v_{i\|jk} - v_{i\|kj} = v_m R_{ijk}^m.$$

We conclude therefore that on a general manifold one cannot interchange the order of the second order covariant derivatives. The difference between the two values is related to the Riemann tensor which represents the "curvature" of the manifold.

3.5   Parallel Transportation of Vectors

To define the concept of "parallel vectors" on a manifold, we observe that in Euclidean space parallel vector has the same angle with any straight line (= geodesic) that connects them. Similarly, we define two vectors in the tangent spaces at two points P, Q on a manifold to be parallel if they make the same angle with the tangent to the geodesic that connects these two points. (Historically, this is referred to as "Levi-Civita transportation")

To see the consequences of this definition, let $\mathbf{v} = v^i \mathbf{x}_i$ be a unit vector in the tangent space at P i.e.

$$g_{ij} v^i v^j = 1,$$

and let G be the geodesic that connects P to Q i.e $\mathbf{x} = (u^1(s), u^2(s))$ where s is the arc-length along the geodesic. The tangent to the geodesic is $\mathbf{w}(s) = \mathbf{x}_i \dot{u}^i$ (where dot represents differentiation with respect to s). Hence, the angle between $\mathbf{v}$ and $\mathbf{w}$ is $g_{ij} v^i \dot{u}^j$. This angle will remain constant along the geodesic if

$$\frac{d}{ds}(g_{ij} v^i \dot{u}^j) = 0,$$

i.e

$$\frac{\partial g_{ij}}{\partial u^k} \dot{u}^k v^i \dot{u}^j + g_{ij} \dot{v}^i \dot{u}^j + g_{ij} v^i \ddot{u}^j = 0.$$

Using the definition of the Christoffel symbols, we have

$$\frac{\partial g_{ij}}{\partial u^k} = \Gamma_{ikj} + \Gamma_{jki}.$$

Therefore,

$$\left[(\Gamma_{ikj} + \Gamma_{jki}) v^i \dot{u}^k + g_{ij} \dot{v}^i \right] \dot{u}^j + g_{ij} v^i \ddot{u}^j = 0.$$

However, using the equation for a geodesic (3.12) and some algebra, we obtain

$$g_{ij}(\dot{v}^i + v^k \Gamma^i_{km} \dot{u}^m) \dot{u}^j = 0. \tag{3.25}$$

Taking the covariant derivative of $g_{ij} v^i v^j = 1$ and using the fact that the covariant derivative of g_{ij} is zero, we obtain

$$g_{ij}(\dot{v}^i + v^k \Gamma^i_{km} \dot{u}^m) v^j = 0. \tag{3.26}$$

Equations (3.25) and (3.26) imply that the vector $(\dot{v}^i + v^k \Gamma^i_{km} \dot{u}^m)$ is orthogonal to the geodesic tangent and to itself but this is impossible (we can choose $\mathbf{v}$ to be non-orthogonal to the tangent); therefore, this expression must be equal to zero viz.

$$\dot{v}^i + v^k \Gamma^i_{km} \dot{u}^m = 0.$$

This equation determines how the components of a vector must change under parallel transportation. For this reason, the Christoffel symbols are referred to as defining a "connection" on the manifold.

3.6 The Torus in R^3

The standard representation of the two-dimensional torus in R^3 is

$$\mathbf{x}(\theta, \phi) \;=\; ((R + a\cos\phi)\cos\theta, (R + a\cos\phi)\sin\theta, a\sin\phi),$$
$$0 \le \phi \le 2\pi,\; 0 \le \theta \le 2\pi. \tag{3.27}$$

where $R > a > 0$ are constants.

In the following, we shall refer to θ and ϕ as u^1, u^2 respectively.

Remark 3.3: An equivalent representation of the torus in cylindrical coordinates (r, θ, z) is

$$(r - R)^2 + z^2 = a^2. \tag{3.28}$$

For a curve $\mathbf{x}(\theta(t), \phi(t))$ on the torus we have

$$\frac{dx}{dt} = -a\sin(\phi)\cos(\theta)\frac{d\phi}{dt} - (R + a\cos(\phi))\sin(\theta)\frac{d\theta}{dt},$$

$$\frac{dy}{dt} = -a\sin(\phi)\sin(\theta)\frac{d\phi}{dt} + (R + a\cos(\phi))\cos(\theta)\frac{d\theta}{dt},$$

$$\frac{dz}{dt} = a\cos(\phi)\frac{d\phi}{dt}.$$

Hence,

$$ds^2 = dx^2 + dy^2 + dz^2 = (R + a\cos(\phi))^2 (d\theta)^2 + a^2 (d\phi)^2.$$

It follows that the components of the torus metric tensor are

$$g_{11} = (R + a\cos(\phi))^2,\quad g_{22} = a^2,\quad g_{12} = g_{21} = 0.$$

Using (3.9) and (3.10), we obtain that the nonzero Christoffel symbols are

$$\Gamma_{112} = -\Gamma_{121} = -\Gamma_{211} = (R + a\cos(\phi))a\sin(\phi). \tag{3.29}$$

$$\Gamma^2_{11} = \frac{(R + a\cos(\phi))\sin(\phi)}{a},\quad \Gamma^1_{12} = \Gamma^1_{21} = -\frac{a\sin(\phi)}{(R + a\cos(\phi)}. \tag{3.30}$$

From these we obtain using (3.24) that the nonzero components of the Riemann tensor are

$$R_{1212} = -R_{1221} = \frac{a\cos(\phi)}{R + a\cos(\phi)},\quad R_{2121} = -R_{2112} = \frac{(R + a\cos(\phi))\cos(\phi)}{a}. \tag{3.31}$$

3.6.1   Geodesics on the Torus

Using (3.12) and (3.30), we derive the following system of nonlinear differential equations for the geodesics on the torus

$$\frac{d^2\theta}{dt^2} + 2\Gamma^1_{12}\frac{d\theta}{dt}\frac{d\phi}{dt} = 0, \tag{3.32}$$

$$\frac{d^2\phi}{dt^2} + \Gamma^2_{11}\left(\frac{d\theta}{dt}\right)^2 = 0. \tag{3.33}$$

It follows that the explicit form of these equations is

$$\frac{d^2\theta}{dt^2} - 2\frac{a\sin(\phi)}{(R+a\cos(\phi)}\frac{d\theta}{dt}\frac{d\phi}{dt} = 0, \tag{3.34}$$

$$\frac{d^2\phi}{dt^2} + \frac{(R+a\cos(\phi))\sin(\phi)}{a}\left(\frac{d\theta}{dt}\right)^2 = 0. \tag{3.35}$$

to simplify these equations, we make the substitution $u = R + a\cos(\phi)$. Since

$$\frac{du}{dt} = -a\sin(\phi)\frac{d\phi}{dt},$$

(3.34) becomes

$$\frac{d^2\theta}{dt^2} + 2\frac{\frac{du}{dt}}{u}\frac{d\theta}{dt} = 0.$$

Hence

$$\frac{\frac{d^2\theta}{dt^2}}{\frac{d\theta}{dt}} = -2\frac{\frac{du}{dt}}{u}.$$

Integrating both sides of this equation with respect to t yields

$$\ln\left(\frac{d\theta}{dt}\right) + 2\ln(u) = c,$$

where c is an arbitrary integration constant. Combining the terms on the left-hand side of this equation and exponentiating we find that

$$\frac{d\theta}{dt} = \frac{C_1}{u^2}, \tag{3.36}$$

where C_1 is an arbitrary constant. Substituting this result in (3.35) leads to

$$\frac{d^2\phi}{dt^2} + \frac{C_1^2\sin(\phi)}{a(R+a\cos(\phi))^3} = 0.$$

To simplify this equation, we multiply it by $\frac{d\phi}{dt}$ and integrate to obtain that

$$\left(\frac{d\phi}{dt}\right)^2 = -\frac{C_1^2}{2a^2(R+a\cos(\phi))^2} + C_2, \tag{3.37}$$

where C_2 is an arbitrary integration constant. It follows then that the nature of the geodesics on the torus will depend on the values of C_1 and C_2, and the different values of these constants will yield geodesics of different nature.

For example, if we let $C_1 = 0$, then $\theta(t)$ is constant and $\phi(t) = C_2 t + C_3$. These geodesics are the "small circles" that loop around the torus limb at constant θ.

Another possibility is to let ϕ be a constant which satisfies (3.37) (with nonzero C_1, C_2). this implies that $\frac{d\theta}{dt}$ is a constant and hence $\theta(t) = At + B$. These geodesics are the circles on the limb of the torus with constant ϕ.

Other more "exotic" types of geodesics exist for other choices of C_1 and C_2.

3.7   Applications to General Relativity

The basic postulate of general relativity is that the distribution of matter determines the geometry of space-time manifold. To determine this relationship, Einstein postulated that

$$E_\nu^\sigma = R_\nu^\sigma - \frac{1}{2}\delta_\nu^\sigma(R - 2\Lambda) = T_\nu^\sigma.$$

where R_ν^σ is the Ricci tensor, R is the curvature, Λ is a constant and T_ν^σ is the energy-momentum tensor of matter. Thus, for a given distribution of energy-momentum, Einstein equation comprises a set of nonlinear partial differential equations which determine the corresponding geometry of space-time..

The "unknown parameter" Λ in these equations has a "history". Initially, it was set to zero. However, Einstein then suggested that it should be included in the equations,but a while latter he conceded that this might be a mistake. However, today the role this constant plays is controversial, and some scientists insist that this parameter is needed for some applications of general relativity.

Today there are many known solution of Einstein equation one of the most (early) important solutions is due to Schwarzschild. This solution corresponds to the geometry of space-time at the exterior of spherical body of mass M. In this region, there is no matter, and therefore, the energy-momentum tensor is zero.

We note that space-times in special and general relativity correspond to a pseudo-Riemannian manifolds since some of the eigenvalues of the metric tensor are negative.

3.7.1   Schwarzchild Solution

Schwarzschild solution to Einstein equations of general relativity (Found in 1916) is an exact solution to these equations. It describes the space-time structure outside a spherical mass. This solution found in many applications, and it was used to confirm the predictions of general relativity regarding the bending of light rays by a star and the perihelic shift of the planet Mercury.

Since the mass under consideration is spherical, it is plausible to assume that the space time metric is of the following form:

$$ds^2 = e^{\nu(r)} dt^2 - \left(e^{\mu(r)} dr^2 + r^2 d\theta^2 + r^2 \sin^2 \theta d\phi^2 \right), \qquad (3.38)$$

(where the speed of light c was scaled to 1). Furthermore, since there is no matter outside the mass, the matter tensor $T^{\mu\nu} = 0$. Hence, the Ricci tensor must satisfy $R_{\mu\nu} = 0$. To derive Einstein equations for the metric (3.38), one has to compute the Christoffel symbols. The nonzero Christoffel symbols for this metric are

$$\Gamma_{11}^1 = \frac{\mu'}{2}, \; \Gamma_{12}^2 = \Gamma_{21}^2 = \Gamma_{13}^3 = \Gamma_{31}^3 = \frac{1}{r},$$

$$\Gamma_{22}^1 = -re^{-\mu}, \; \Gamma_{33}^1 = -re^{-\mu} \sin^2 \theta, \; \Gamma_{14}^4 = \Gamma_{41}^4 = \frac{\nu'}{2},$$

$$\Gamma_{23}^3 = \Gamma_{32}^3 = \cot \theta, \; \Gamma_{33}^2 = -\sin\theta\cos\theta, \; \Gamma_{44}^1 = \frac{\nu'}{2} e^{\nu - \mu}.$$

Using this data we find that the nonzero components of $R_{\mu\nu} = 0$ are four:

$$R_{11} = \frac{1}{2}\nu'' - \frac{\mu'}{r} - \frac{1}{4}\mu'\nu' + \frac{1}{4}(\nu')^2 = 0, \qquad (3.39)$$

$$R_{33} = \sin^2\theta R_{22} = \sin^2\theta \left(e^{-\mu} + \frac{1}{2}re^{-\mu}(\nu' - \mu') - 1 \right) = 0, \qquad (3.40)$$

$$R_{44} = e^{\nu - \mu} \left(\frac{1}{2}\nu'' - \frac{\mu'}{r} - \frac{1}{4}\mu'\nu' + \frac{1}{4}(\nu')^2 \right), \qquad (3.41)$$

where primes denote differentiation with respect to r. From these equations, we infer that $R_{11} + e^{\mu - \nu} R_{44} = 0$ will be satisfied if $\nu' + \mu' = 0$. Hence $\mu = -\nu$. From (3.40), we then obtain

$$e^\nu (1 + r\nu') = 1,$$

Hence

$$e^\nu = e^{-\mu} = 1 - \frac{\alpha}{r},$$

where α is an integration constant. The determination of this constant can be made by asymptotic reduction to Newtonian theory, and it turns out that

$\alpha = \frac{2GM}{c^2} = r_s$. This is usually referred to as the Schwarzschild radius of the star. For the Sun $r_s \approx 3km$.

Thus the Schwarzschild metric tensor outside a spherical mass is

$$ds^2 = (1 - \frac{r_s}{r})dt^2 - \left[(1 - \frac{r_s}{r})^{-1}dr^2 + r^2 d\theta^2 + r^2 \sin^2 \theta d\phi^2\right]. \qquad (3.42)$$

We note that space-times in Special and general relativity correspond to a pseudo-Riemannian manifolds since some of the eigenvalues of the metric tensor are negative.

The attached Maple program "Sch.mw" verifies that this metric tensor satisfies Einstein equations with $\Lambda = 0$.

Appendix 3A: Light Ray Bending by a Star

Based on overwhelming experimental evidence light rays in free space were thought to be propagating in straight lines. One of the major predictions of Einstein theory of general relativity was that light rays passing in the vicinity of a star should be bended. The experimental evidence for this fact was obtained by Eddington during the 1919 eclipse of the Sun. This experimental result provided strong evidence for theory of General Relativity and it acceptance as a new major physical theory (Figure 3.1).

According to the theory of special relativity, the trajectory of a light ray in free space is characterized by its null line element

$$ds^2 = t^2 - x^2 - y^2 - z^2 = 0,$$

where the value of c was normalized to 1 Similarly, according to general relativity, light rays travel along null-geodesics. We note that for null geodesics $ds^2 = 0$, and therefore, one has to use a different parameterization for these geodesics.

The resulting equations for these geodesics are:

$$\frac{d}{dw}\left\{\left(1 - \frac{r_s}{r}\right)\frac{dt}{dw}\right\} = 0, \qquad (3A.1)$$

$$\frac{d}{dw}\left(r^2 \frac{d\theta}{dw}\right) - r^2 \sin\theta \cos\theta \left(\frac{d\phi}{dw}\right)^2 = 0, \qquad (3A.2)$$

$$\frac{d}{dw}\left(r^2 \sin^2\theta \frac{d\phi}{dw}\right) = 0, \qquad (3A.3)$$

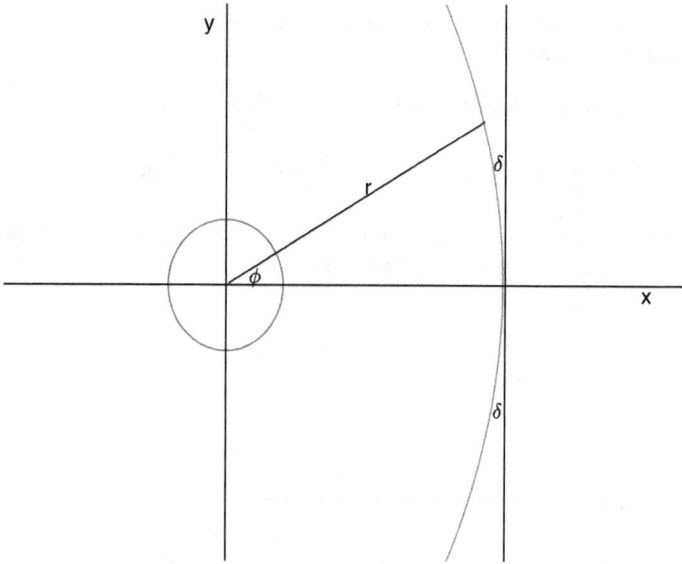

FIGURE 3.1
Light bending by the gravitational field of a star.

$$0 = \left(1 - \frac{r_s}{r}\right)\left(\frac{dt}{dw}\right)^2 - \left(1 - \frac{r_s}{r}\right)^{-1}\left(\frac{dr}{dw}\right)^2 + r^2\left(\frac{d\theta}{dw}\right)^2 + r^2\sin^2\theta\left(\frac{d\phi}{dw}\right)^2.$$
$$(3A.4)$$

By proper orientation of the coordinate system, we can align the path of the light ray far away from the star along a geodesic in a plane defined by

$$\theta = \frac{\pi}{2}, \quad \frac{d\theta}{dw} = 0.$$

then (3A.3) and (3A.1) yield

$$r^2\frac{d\phi}{dw} = h_1,$$

$$\left(1 - \frac{r_s}{r}\right)\frac{dt}{dw} = h_2,$$

where h_1 and h_2 are constants. Substituting these results in (3A.4) yields

$$0 = \left(1 - \frac{r_s}{r}\right)^{-1}h_2^2 - \left(1 - \frac{r_s}{r}\right)^{-1}r^2 - \frac{h_1}{r^2}.$$

Substituting $v(\phi) = \frac{1}{r(\phi)}$ in (3A.4) and differentiating the result yields the following,

$$\frac{dv}{d\phi}\left(\frac{d^2v}{d\phi^2} + v - \frac{3r_s}{2}v^2\right) = 0.$$

Assuming v is not a constant we obtain the following differential equation for v

$$\frac{d^2v}{d\phi^2} + v = \epsilon v^2, \quad v(0) = \frac{1}{r_0}, \tag{3A.5}$$

where $\epsilon = \frac{3r_s}{2}$. Since ϵ is a small parameter, it is appropriate to solve this equation by a perturbation expansion in ϵ

$$v(\phi, \epsilon)v_0(\phi) + \epsilon v_1(\phi) + \cdots$$

To first order in ϵ we have

$$\epsilon^0 : \frac{d^2v_0}{d\phi^2} + v_0 = 0, \quad v_0(0) = \frac{1}{r_0},$$

$$\epsilon^1 : \frac{d^2v_1}{d\phi^2} + v_1 = v_0^2, \quad v_1(0) = 0.$$

Solving these equations for v_0, v_1 we find that

$$v = \frac{\sin\phi}{r_0} + \frac{\epsilon}{2r_0^2}\left(1 + \frac{1}{3}\cos(2\phi)\right) + \cdots$$

The first term in this equation represents a straight line trajectory, whereas the second term represents the deviation from this trajectory. Asymptotically (viz large distances from the star) $v = 0$ and $0 < \phi \ll 1$ (or $\pi - \phi \ll 1$). Hence, we can approximate $\sin(\phi) = \delta$ and $\cos(2\phi)$) by 1. Therefore,

$$\frac{\delta}{r_0} + \frac{2\epsilon}{3r_0^2} = 0.$$

Thus, the deviation angle δ from $\phi = 0$ to $\phi = \frac{\pi}{2}$ is

$$\delta = \frac{-2\epsilon}{3r_0}.$$

The minus sign represents the fact that the light ray is bended toward the star. However, the total deflection of the light ray is given by the angle between the two asymptotes of the trajectory at $\pm\infty$. Therefore, due to the symmetry of the trajectory with respect to the x-axis, the total deviation of the ray from a straight line is 2δ.

Appendix 3B: Maple and MATLAB® Programs

With this chapter, we provide the following maple programs:

sphere.mw: program to compute the Christoffel symbols and Riemann, Ricci tensors and curvature for the 2-d sphere

torus.mw: program to compute the Christoffel symbols, Riemann and Ricci tensors and curvature for the 2-d Torus

Riemann3.mw: program to compute the Christoffel symbols, Riemann and Ricci tensors and curvature for the 2-d cylinder

Sch.mw: program to verify that Schwarzchild solution satisfies Einstein General relativity equations.

Exercises

1. Compute the Christoffel symbols for the three-dimensional ellipsoid
$$\mathbf{x} = (a \sin \psi \sin \phi \cos \theta, b \sin \psi \sin \phi \sin \theta, c \sin \psi \cos \phi, d \cos \psi)$$

2. Compute the Christoffel symbols for the three-dimensional hyperboloid of one sheet
$$\mathbf{x} = (a \cosh(u) \sin(\phi) \cos(\theta), b \cosh(u) \sin(\phi) \sin(\theta), c \cosh(u) \cos(\phi), d \sinh(u)).$$

3. Compute the geodesics on the three-dimensional sphere in R^4.
   The equation of the sphere in "spherical coordinates" is
$$\mathbf{x} = (R \sin \psi \sin \phi \cos \theta, R \sin \psi \sin \phi \sin \theta, R \sin \psi \cos \phi, R \cos \psi)$$

4. Compute the geodesics on the three-dimensional ellipsoid
$$\mathbf{x} = (a \sin \psi \sin \phi \cos \theta, b \sin \psi \sin \phi \sin \theta, c \sin \psi \cos \phi, d \cos \psi)$$

5. Compute the geodesics on the two-dimensional Torus in R^3
$$\mathbf{x}(\theta, \phi) = ((a{+}r \cos \theta) \cos \phi, (a{+}r \cos \theta) \sin \phi, r \sin \theta), \quad 0 \le \phi \le 2\pi, \ 0 \le \theta \le 2\pi$$

Further Readings

R. Adler, M. Bazin, M. Schiffer (1975), *Introduction to General Relativity*, 2nd edition, McGraw-Hill, New York.

E. Kreyszig (1991), *Differential Geometry*, Dover, New York.

G. C. Mcvittie (1965), *General Relativity and Cosmology*, University of Illinois Press, Champaign, IL.

I. L. Shapiro (2019), *A Premier in Tensor Analysis and Relativity*, Springer, Berlin, Germany.

I. S. Sokolnikoff (1965), *Tensor Analysis*, Wiley, Hoboken, NJ.

K. I. Throne, R. D. Blandford (2021), *Relativity and Cosmology*, Princeton University Press, Princeton, NJ.

4

Basic Topology and Analysis

4.1 Basic Notions of Topology

Definition 4.1: A topology on a set S consists of a collection of subsets of S $\{I_\alpha\}$ with the following properties:

1. The union of any number of sets $I_p \in \{I_\alpha\}$ is a set in $\{I_\alpha\}$
2. The intersection of any finite number of sets in $\{I_\alpha\}$ is a set in $\{I_\alpha\}$.
3. S and the empty set are in $\{I_\alpha\}$

Such a set S is referred to as a **topological space**.

Remark 4.1

1. A set can have several (completely different) topologies.
2. The intersection of infinite number of sets in $\{I_\alpha\}$ does not have to be in $\{I_\alpha\}$

We shall refer to the sets in $\{I_\alpha\}$ as open sets

Example 4.1: Consider the real line R. A topology on R consists of all open intervals of the form (a, b) (viz. the intervals on R from a to b without the end points) their unions and finite intersections.

Let $I_n = (-\frac{1}{n}, 1 + \frac{1}{n})$, $n = 1, \ldots \infty$ It is obvious that each I_n is open but the intersection of all these open sets is the interval $[0, 1]$ which is not open.

Definition 4.2: A **neighborhood** of a point $P \in S$ is a set in I_α which contains P.

Definition 4.3: A set $D \in S$ is an **open set** if each of the points in D has a neighborhood which is contained (completely) in D.

Definition 4.4: A set whose complement in S is open is referred to as **closed set**.

DOI: 10.1201/9781003587422-4

Example 4.2: The unit circle $C = \{(x,y)|x^2 + y^2 < 1\}$ in R^2 (with the usual topology) is open. On the other hand, $D = \{(x,y)|x^2 + y^2 \leq 1\}$ is closed since the points on the boundary of D do not have an open neighborhood which is contained completely in D. Furthermore the complement of D in R^2 is open.

Definition 4.5: The **Cartesian product** $A \times B$ of two sets A, B is the set of all points (a,b) where $a \in A$ and $b \in B$

If A, B are topological spaces then the corresponding topology on $A \times B$ consists of all sets of the form $C \times D$ where C, D are open sets in A, B respectively.

4.2 Basic Notions from Analysis

Definition 4.6: Let S, T be topological spaces. A function $f : S \to T$ is **continuous** at a point s if for any neighborhood of V of $f(s)$ there is a neighborhood U of s so that $f(U) \subset V$

Example 4.3: How this definition corresponds to the usual definition of continuity on the real line?

Let f be a function $R \to R$. If the function is continuous at s. The definition of continuity states that for "any neighborhood V of $f(s)$". Such a neighborhood is of the form of an open interval around $f(s)$ viz. $|f(x) - f(s)| \leq \epsilon$. The function is continuous at s if for any such neighborhood of $f(s)$ there exists a neighborhood U of s i.e $|x - s| < \delta$ so that for each $x \in U$, $f(U) \subset V$.

Thus, we regain the usual formulation of continuity: For any $\epsilon > 0$ there exist $\delta > 0$ so that for all $|s - x| < \delta$, $|f(x) - f(s)| < \epsilon$.

Definition 4.7: A function f is "continuous" if it is continuous at each point of its domain.

Definition 4.8: A mapping $f : A \to B$ so that $f(a_1) = f(a_2)$ implies that $a_1 = a_2$ is called **injection** (or one-to-one map).

In this case f^{-1} exists.

Note that the domain of f does not have to be all of A.

Definition 4.9: If the range of $f : A \to B$ is all of B, we then say that f is a **surjection**

Definition 4.10: An injection $f : A \to B$ whose domain is A and range is B is called a **bijective map**.

Definition 4.11: A function $f : S \to T$ is a homeomorphism if it an injection which is continuous and maps open sets in S to open sets in T. If f is bijective and f^{-1} is continuous we say that S, T are **homeomorphic**.

Remark 4.2: Note that this concept is NOT same as algebraic "homomorphism"

Example 4.4: Show that the topological spaces $(0, 1)$ and $(1, \infty)$ with the usual topology on R are homeomorphic

Solution: let $f(0, 1) \to (1, \infty)$ be defined as

$$f(x) = \frac{1}{x}.$$

This function is obviously an injection. It is continuous and maps open sets to open sets. It is also surjective and f^{-1} is continuous. Therefore these two sets are homeomorphic.

Example 4.5: A circle and a square in R^2 are homeomorphic since there exists a continuous map that takes each point on the circle to a unique point on the square.

Example 4.6: Let $f : [0, 2\pi) \to S^1$ where S^1 is the unit circle in the plane (both spaces with their usual subspace topology). Show that $f(t) = (\cos t, \sin t)$ is **not** a homeomorphism. Obviously f is bijective and continuous however the inverse of f is not continuous since a neighborhood of $(1, 0) \in S^1$ is mapped by f^{-1} to points in the neighborhoods of 0 and 2π.

Remark 4.3: There are several types of functions $f : R^k \to R^m$

1. f is a C^n function (where n is an integer) if it is differentiable n times and the nth derivative is continuous. For example, $f \in C^0$ if it is continuous.

2. $f \in C^\infty$ if it is differentiable to any order. Such functions are referred to as "smooth functions". For example, $\sin(x)$ is a C^∞ function.

3. f is analytic if it is a C^∞ function and the Taylor expansion of f converges to f.

Observe that not every C^∞ function is analytic.

Example 4.7: Cauchy function

$$f(x) = \exp\left(\frac{-1}{x^2}\right),$$

is a C^∞ function but is not analytic due to the fact that all its derivatives at $x = 0$ are zero, and therefore, the Taylor expansion of f around $x = 0$ is identically zero, but the function is zero only at $x = 0$.

For complex functions $f : C \to C$, if f is $C^1(C)$, then it is analytic.

Definition 4.12: An injection $f : R^n \to R^m$ where f and f^{-1} are C^∞ functions is referred to as a **diffeomorphism**. We observe that the domain of f must be an open set in R^n. (This follows from the definition of injection)

Example 4.8: The mapping $f : R^n \to R^n$ which is defined by $f(\mathbf{x}) = \mathbf{x} + \mathbf{a}$ (viz. translation by $\mathbf{a}$) is a diffeomorphism

Example 4.9: The mapping $f : R^2 \to R^2$ which is defined as $(x, y) \to (u, v)$ where

$$u = x\cos\theta + y\sin\theta, v = -x\sin\theta + y\cos\theta, \ 0 \le \theta < 2\pi,$$

(rotation by angle θ) is a diffeomorphism

This definition of diffeomorphism can be generalized to one between differential manifolds:

Definition 4.13: Let M, N be smooth differentiable manifolds. A differentiable map $f : M \to N$ is a diffeomorphism if it is a bijection and its inverse $f^{-1} : N \to M$ is differentiable.

4.2.1 Properties of Diffeomorphism

1. If $f : M_1 \to M_2$ and g:$f : M_2 \to M_3$ are diffeomorphisms then $g \circ f$ is a diffeomorphism $M_1 \to M_3$

2. If f is a diffeomorphism, then the inverse mapping f^{-1} is also a diffeomorphism.

Definition 4.14: The **Jacobian** of a C^∞ mapping $f : R^n \to R^m$ is the $m \times n$ matrix

$$J(\mathbf{x}) = \left[\frac{\partial f_i}{\partial x_j}(\mathbf{x}) \right], \ i = 1, \ldots, m, \ j = 1, \ldots, n.$$

Theorem 4.1: Inverse Function Theorem If the Jacobian of a smooth map f is nonsingular at a point then f has a smooth inverse when restricted to some neighborhood of that point.

Example 4.10: Determine if the mapping $f : R^2 \to R^2$ where

$$f(x, y) = (x^3, y^2),$$

is a diffeomorphism.

Answer: No. It is not a diffeomorphism since $f(x, y) = f(x, -y)$ which means that f is not injective.

Example 4.11: Determine if the mapping $f : R^2 \to R^2$ where

$$f(x, y) = (\cos(x^2 + y^2), \sin(x^2 + y^2)),$$

is a diffeomorphism .

Answer No. The image of R^2 under f is the unit circle.

Example 4.12: Determine if $f : R \to R$ where $f(x) = x^3$ is a diffeomorphism.

Answer: Although the mapping is bijective its derivative at zero is zero. It follows that its inverse $g = x^{1/3}$ is not differentiable at this point and therefore f is not a diffeomorphism. However, f is a homeomorphism $R \to R$.

4.3   Summary

1. Homeomorphism is a mapping $f : S \to T$ between topological space with the following properties:

   - f is an injection (one-to-one)
   - f maps open sets to open sets
   - f and f^{-1} are continuous

2. We say that S, T are are homeomorphic if f is a bijective homeomorphism.

3. $f : R^n \to R^m$ is a diffeomorphism if it is homeomorphism and

   - domain of f is an open set
   - f and f^{-1} are C^∞ functions.

   To show that a mapping $f : R^m \to R^n$ (or more generally between two differential manifolds) is a diffeomorphism one has to show that it is bijective (viz. one-to-one and onto), smooth, and the inverse map is also smooth.

Exercises

1. Let $S = \{\mathbf{x} \in R^n, |\mathbf{x} - \mathbf{x_0}| < r\}$. Show that S is open. Hint: Consider the function $f(\mathbf{x}) = |\mathbf{x} - \mathbf{x_0}|$.

2. Let A be a subset of R^n and B is a closed subset of R^n so that $A \subset B \subset \bar{A}$ (where $\bar{A}$ is the closure of A) Show that $B = \bar{A}$.

   Remark 4.4: The closure of A consists of all points **b** for which there exist a sequence $\mathbf{a_n} \in A$ so that $\lim \mathbf{a_n} = \mathbf{b}$.

3. Let S be the unit circle in R^2 and A be the subset of S which consists of $(\cos(\theta), \sin(\theta))$ $0 < \theta < \pi$. Show that the mapping $(\cos(\theta), \sin(\theta)) \to \theta$ is an injection to an open subset of R.

4. The set $S = (\sin(2\theta, \sin\theta)$ represent the figure eight in R^2. Show that the mapping $S \to R$

$$f((\sin(2\theta), \sin\theta)) = \theta, \ 0 < \theta < 2\pi$$

is an injection to the interval $(0, 2\pi)$.

Further Readings

F. Brickell , R. S. Clark (1970), *Differential Manifolds*, Van Nostrand.

G. Gross, E. Meinrenken (2023), *Manifolds, Vector Fields and Differential Forms*, Springer, Berlin, Germany.

S. Lovett (2020), *Differential Geometry of Manifolds*, CRC Press, Boca Raton, FL.

J. R. Munkres (1997), *Analysis on Manifolds*, CRC Press, Boca Raton, FL.

T. B. Singh (2019), *Introduction to Topology*, Springer, Berlin, Germany.

M. Spivak (1965), *Calculus on Manifolds*, W.A Benjamin, California,.

5

Differential Manifolds

5.1 Introduction

The roots of differential geometry are grounded in the theory and applications of curves and surfaces in R^3 which can be generalized to R^n. However, it turned out that other mathematical entities which do not belong to this class of geometrical objects have somewhat similar "structure". This has led to the modern more abstract concept of "Differential Manifolds" which is the subject of this chapter.

5.2 Charts and Atlases

Let M be a topological space.

Definition 5.1: A homeomorphism $\mathbf{x} : M \to R^n$ whose range is an open set in R^n is called a chart on M. (remember the domain of such a homeomorphism does not have to be all of M). If $m \in M$ is in the domain of $\mathbf{x}$, then

$$\mathbf{x}(m) = (x^1(m), \ldots, x^n(m)),$$

represents the coordinates of the point m in the chart.

In other words, the mapping $\mathbf{x}$ has to be an injection into R^n whose domain and range are open sets and maps open sets in its domain to open sets in R^n.

Example 5.1: Let M be the set of all $m \times n$ matrices with real entries. The mapping $M \to R^{mn}$ given by

$$[a_{ij}] \to (a_{11}, \ldots, a_{1,n}, a_{21}, \ldots, a_{2n} \ldots),$$

is a (global) chart for M since it is obviously an injection and R^{mn} is open.

However, in general, one cannot find global charts.

DOI: 10.1201/9781003587422-5

Example 5.2: Let $M = S^1$ the unit circle in R^2. Let $A \subset S^1$ be the set

$$A = \{(\cos(2\pi t), \sin(2\pi t)), 0 < t < 1\}.$$

The mapping $x : (\cos(2\pi t), \sin(2\pi t)) \to t$ is a chart on A and t is the coordinate assigned to this point. Observe that in this case the chart cannot be "extended" to all of M. The domain of this chart consists of all points in S^1 except $(1, 0)$. ("pinched" circle).

Let $B \subset S^1$ be the set

$$B = \{(\cos(2\pi t), \sin(2\pi t)), -1/2 < t < 1/2\}.$$

The mapping $y : (\cos(2\pi t), \sin(2\pi t)) \to t$ is a chart on B. Once again this chart cannot be extended to all of S^1. The domain of this chart consists of all points in S^1 except $(-1, 0)$. ("pinched" circle).

Definition 5.2: A collection of n-dimensional charts that covers M is called an **Atlas**. An atlas is C^∞ atlas on M if for any two charts x, y in the atlas whose domain intersect

$$y \circ x^{-1} : R^n \to R^n,$$

is a diffeomorphism **on the intersection of their domains** (Figure 5.1).

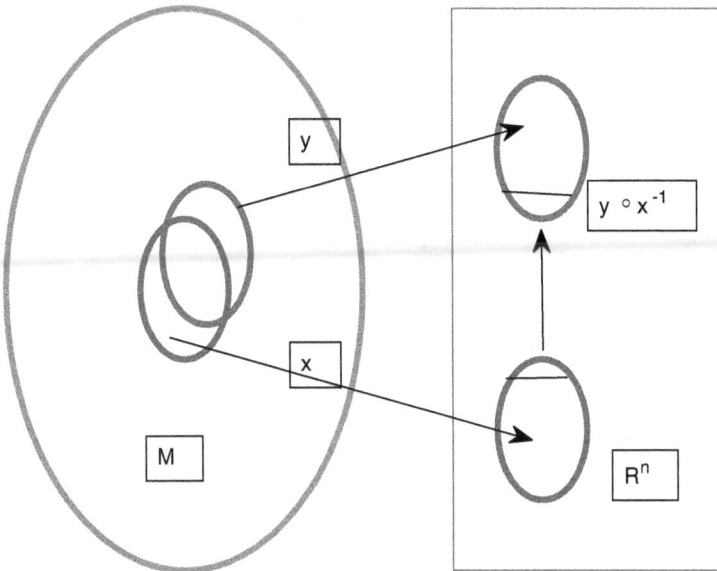

FIGURE 5.1
$y \circ x^{-1}$ is a diffeomorphism on the domain intersection of the two chart.

Example 5.3: The two charts in Examples 5.1 and 5.2 cover S^1 together and they form C^∞ atlas on S^1. The intersection of their domains consists of S^1 except the two points $(1,0)$ and $(-1,0)$. For $0 < t < 1/2$ the mapping $y \circ x^{-1}$ is the identity mapping on the open interval $(0, \frac{1}{2})$. For $1/2 < t < 1$ the mapping $y \circ x^{-1}$ is given by $s \to s - 1$. This shows that $y \circ x^{-1}$ is a diffeomorphism $R \to R$ on the intersection of their domain.

Example 5.4: Let

$$M = \{(t,0) \in R^2, -1 < t < 1\} \cup \{(t,t) \in R^2, 0 < t < 1\}.$$

On this set the mappings

$$x : \{(t,0) \in R^2, -1 < t < 1\} \to t,$$

$$y : \{(t,0) \in R^2, -1 < t < 0\} \cup \{(t,t) \in R^2, 0 < t < 1\} \to t.$$

are charts that together cover M. However, they do not form a C^∞ atlas on M. In fact, although $y \circ x^{-1}$ is the identity function its domain $(-1,0)$ is not an open set. Hence $y \circ x^{-1}$ is not a diffeomorphism.

In general, a set M can have more than one atlas. Two such atlases are **equivalent** if the their union is a C^∞ atlas on M.

Example 5.5: Let

$$M = \{(\sin 2t, \sin t) \in R^2, \ 0 < t < 2\pi\}.$$

This is the Figure 8.

Consider the following two charts which trace M in opposite directions (Figure 5.2).

$$x : (\sin 2t, \sin t) \to t, \ 0 < t < 2\pi,$$

$$y : (\sin 2t, \sin t) \to t, \ -\pi < t < \pi.$$

These are two global charts of M. Hence each is a an atlas for M. However these two atlases are not equivalent since $y \circ x^{-1}$ is mapping of $(0, 2\pi)$ to $(-\pi, \pi)$ with the following values:

$$t \to t, \text{ for } 0 < t < \pi, \ t \to 0 \text{ for } t = \pi, \ t \to t - 2\pi \text{ for } \pi < t < 2\pi,$$

which is not continuous. A plot of $y \circ x^{-1}$ is of the following form:

Definition 5.3: An atlas on M is complete if it is not contained in any other atlas.

Theorem 5.1: A C^∞ atlas on M is contained in just one complete C^∞ atlas.

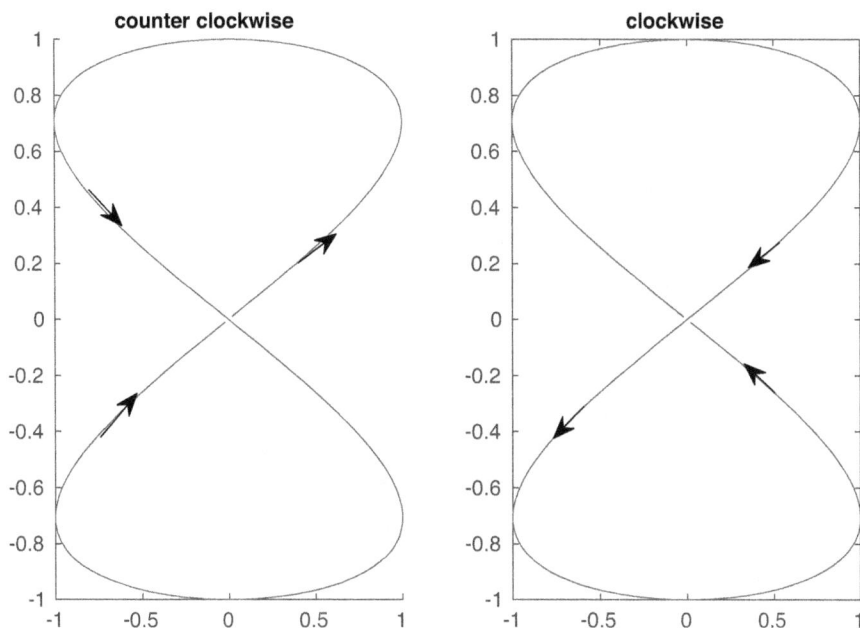

FIGURE 5.2
Two different charts of the Figure 8.

Definition 5.4: A **differential manifold** of dimension n is a topological space M with a complete C^∞ atlas into R^n. In the following we refer to this structure as a "manifold".

Intuitively, a manifold is a space that in the neighborhood of any of it points "looks like" an open set R^n. However, it global structure might be very different than R^n (Figure 5.3).

There are two methods to construct manifold in an "easy way".

1. If M, M' are manifolds then their Cartesian product $N = M \times M'$ is also a manifold. The charts on N are simply the Cartesian products of the charts on M and M'.

Example 5.6: We already showed that the unit circle in R^2 is a manifold. Hence the n-torus which is defined as

$$T^n = S^1 \times S^1 \ldots \times S^1, \quad \text{(Cartesian product of } n \text{ circles)}$$

is also a manifold.

2. The second method follows from the following theorem:

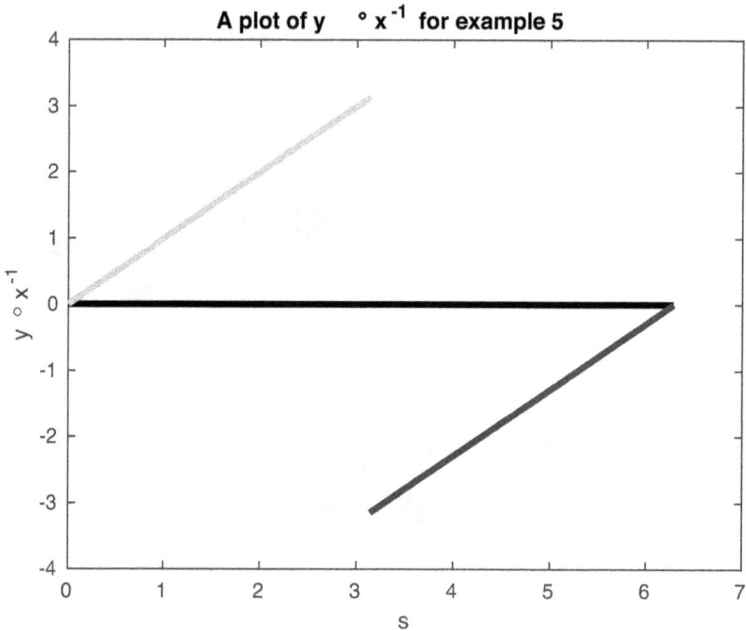

FIGURE 5.3
A plot of $y \circ x^{-1}$ for Example 5.5.

Theorem 5.2: Let $f : R^n \to R$ be a differentiable function and $M = f^{-1}(0)$. If at each point $\mathbf{x} \in M$ the matrix

$$J(\mathbf{x}) = \left(\frac{\partial \mathbf{f}}{\partial \mathbf{x_i}}(\mathbf{x}) \right),$$

has rank one then one can induce a manifold structure of dimension $n-1$ on M.

Example 5.7: The unit circle S^1 is the locus of $f^{-1}(0)$ where $f(x,y) = x^2 + y^2 - 1$. The matrix $J = (2x, 2y)$ and has rank one on the circle. Therefore S^1 is a manifold. The induced atlas is defined on the sets $\{U_1, U_2, U_3, U_4\}$ where

$$U_1 = \{(x,y) \in S^1, x > 0\}, \quad U_2 = \{(x,y) \in S^1, y > 0\},$$

$$U_3 = \{(x,y) \in S^1, x < 0\}, \quad U_4 = \{(x,y) \in S^1, y < 0\},$$

and corresponding charts are

$$z_1 : U_1 \to (0,1), \ z_1(x,y) = x, \ z_2 : U_2 \to (0,1), \ z_2(x,y) = y,$$

$$z_3 : U_3 \to (-1,0), \ z_1(x,y) = x, \ z_4 : U_4 \to (-1,0), \ z_2(x,y) = y.$$

One can show that this atlas is equivalent to one defined before on S^1. Hence, both atlases define the same differential structure on S^1.

Similarly, the unit n-dimensional sphere is defined as $f^{-1}(0)$ where

$$f(x_1, \ldots, x_{n+1}) = x_1^2 + \cdots + x_{n+1}^2 - 1.$$

The induced atlas is defined by the $2(n+1)$ sets

$$U_i = \{(\mathbf{x}) \in S^n, x_i > 0\}, \quad U_{i+n+1} = \{(\mathbf{x}) \in S^n, x_i < 0\}, \quad i = 1, \ldots, (n+1),$$

and the corresponding charts are

$$z_i : U_i \to (0,1), \ z_i(\mathbf{x}) = x_i, \ z_{i+n+1} : U_{i+n+1} \to (-1,0), \ z_i(\mathbf{x}) = x_i.$$

Example 5.8: Another atlas on S^n can be defined by stereographic projections from the points $\mathbf{v}_1 = (0, \ldots, 0, 1)$ and $\mathbf{v}_2 = (0, \ldots, -1)$ onto the plane $x_{n+1} = 0$ in R^{n+1}. These mappings define charts on the punctured sphere

$$U_1 = \{S^n - \mathbf{v}_1\}, \quad U_2 = \{S^n - \mathbf{v}_2\},$$

by

$$\mathbf{z}_1 : U_1 \to R^n, \ \mathbf{z}_2 : U_2 \to R^n,$$

where

$$\mathbf{z}_1(\mathbf{x}) = \frac{x_i}{1 - x_{n+1}}, \ \mathbf{z}_2(\mathbf{x}) = \frac{x_i}{1 + x_{n+1}}, \ i = 1, \ldots, n.$$

It can be shown that this atlas is equivalent to the previous atlas that was defined on S^n.

5.2.1 Orientation

Definition 5.5: Let M be a (smooth) manifold. An atlas of M is oriented if for all charts (ϕ_α, U_α) and (ϕ_β, U_β) in the atlas with coordinate transformation $\psi_{\alpha\beta} : \phi_\beta \phi_\alpha^{-1}$ on $U_\alpha \cap U_\beta$ the determinant of $J(\psi_{\alpha\beta})$ is positive on its domain.

Definition 5.6: A manifold M is orientable if it possesses an oriented atlas.

5.3 Differentiable Mappings

Let M, N be two differentiable manifolds. We want to define the concept of a differentiable mapping $f : M \to N$.

Let m a point in the domain of f and let x, y be charts on neighborhoods U, V of m and $f(m)$ respectively. We shall call

$$F = y \circ f \circ x^{-1} : R^n \to R^{n'},$$

a coordinate representative of f. We shall say that f is differentiable at m if F is C^∞ at $x(m)$. To prove that this definition "makes sense" we must show that it is independent of the choice of the charts at m and $x(m)$.

Let x', y' be other charts at m and $x(m)$ on neighborhoods U', V' respectively. Using these charts, the representative of f will be

$$G = y' \circ f \circ (x')^{-1} : R^n \to R^{n'}.$$

To show that G is C^∞ at $x(m)$ we observe that on $U \cap U'$ and $V \cap V'$ we have

$$G = (y' \circ y^{-1}) \circ F \circ (x \circ (x')^{-1}).$$

Hence if F is C^∞ at at $x(m)$ so is G.

Definition 5.7: We shall say that $f : M \to N$ is differentiable if it is differentiable at every point of its domain.

Example 5.9: the determinant function

$$\det : M(n \times n, R) \to R,$$

has a polynomial representation using the standard charts on these manifolds and therefore is differentiable.

Definition 5.8: a diffeomorfism $f : M \to N$ is an injection such that f and f^{-1} are differentiable. We say that M, N are diffeomorfic if there is a global diffeomorfism from M to N.

Example 5.10: Show that the mapping

$$f : R^2 \to R^2, \quad f(x, y) = (xe^y + y, xe^y - y),$$

is a diffeomorphism.

Solution: If (u,v) are the image points of (x,y), then

$$u = xe^y + y, \quad v = xe^y - y,$$

and therefore the inverse map is

$$x = \frac{u + v}{2e^{(u-v)/2}}, \quad y = \frac{u - v}{2},$$

is one to one. Furthermore, both f and f^{-1} are C^∞ functions. We conclude that f is a diffeomorphism.

Example 5.11: Let $f : R^3 \to R^3$ be the mapping

$$f(x, y, z) = (x \cos z - y \sin z, x \sin z + y \cos z, z).$$

Show that the restriction of f to the unit sphere S^2 is a diffeomorphism of S^2 to itself.

Solution: If (u,v,w) are the image points of (x,y,z), then it is easy to see that

$$x = u \cos w + v \sin w, \quad y = -u \sin w + v \cos w, \quad z = w.$$

Hence on S^2 the mapping is one-to one and both f and f^{-1} are C^∞ functions. We conclude that f is a diffeomorphism.

5.3.1 The Riemann Sphere-Stereographic Mappings

Consider the unit $2-d$ sphere in R^3 which is centered at the origin viz. $(0,0,0)$. The north pole on this sphere is located at $N = (0,0,1)$ and the points on the sphere satisfy,

$$x^2 + y^2 + z^2 = 1.$$

We now define a mapping of the points on sphere to $x - y$ plane as follow: Let $P = (x_1, y_1, z_1)$. point on the sphere. Draw a straight line which contains N and P. The intersection point (u,v,0) of this line with the $x - y$ plane define a stereographic mapping T_1 from the sphere to the $x - y$ plane. Thus,

$$T_1(P) = T_1((x_1, y_1, z_1)) = (u_1, v_1, 0).$$

To obtain an explicit representation of the transformation T, we recall that a parametric represent ion of a line in R^3 is

$$x = a_1 + b_1 t, \quad y = a_2 + b_2 t, \quad z = a_3 + b_3 t,$$

where $a_i, b_i, i = 1, 2, 3$ are constants. To determine these constants, we let $t = 0$ correspond to the point N and $t = 1$ correspond to the point $(u, v, 0)$. Using this data to solve for the coefficients $a_i, , b_i$, we find that the equation this line is

$$x = u_1 t, \quad y = v_1 t, \quad z = 1 - t.$$

To determinate the value of the parameter t which correspond to the point P, we use the fact that this point is on this line and on the sphere. Therefore, the value of the parameter t that correspond to this point must satisfy

$$(u_1 t)^2 + (v_1 t)^2 + (1 - t)^2 = 1.$$

Solving this equation for t, we have

$$t_1 = \frac{2}{1 + u_1^2 + v_1^2}.$$

It follows then that

$$x_1 = u_1 t_1, \quad y_1 = v_1 t_1, \quad z_1 = \frac{u_1^2 + v_1^2 - 1}{1 + u_1^2 + v_1^2}.$$

However, since $t_1 = 1 - z_1$, we can express these results as

$$u_1 = \frac{x_1}{1 - z_1}, \quad v_1 = \frac{y_1}{1 - z_1}.$$

The explicit representation of the transformation T_1 is

$$T_1(x_1, y_1, z_1) = \left(\frac{x_1}{1 - z_1}, \frac{y_1}{1 - z_1}, 0 \right). \tag{5.1}$$

We observe that under this transformation the point N is mapped to the point of infinity in the $x - y$ plane. A neighborhood of the point N on the sphere consists of the interior of a circle around this point. This neighborhood is mapped under the streographic projection to the exterior of a circle around the origin is $x - y$ plane. Hence, it is "legitimate" to say that a neighborhood of the point at infinity in the plane consists of the exterior of a circle around the origin in this plane.

We note that another stereographic mapping of the sphere can be obtained by replacing the point $N = (0,0,1)$ ("north pole") by $M = (0,0,-1)$ ("south pole"). If we use M to induce the stereographic mapping, we shall have

$$(u_2, v_2, 0) = T_2(x_1, y_1, z_1) = \left(\frac{x_1}{1 + z_1}, \frac{y_1}{1 + z_1}, 0 \right). \tag{5.2}$$

5.3.1.1 Stereographic Atlas

The two-dimensional sphere can be represented by the Cartesian product of $S^1 \times S^1$ where S^1 is a circle in two dimensions. Therefore, it is a manifold with charts which are the Cartesian products of those for S^1. The stereographic mapping provides an atlas with two charts. The first consists of the "pinched sphere" without the point N and the other chart is S^2 without the point M together these two chart cover S^2. To show that on the intersection of these two charts $T_1 \circ T_2^{-1}$ is a diffeomorphism ,we note using (10.2) and (1.9) that

$$\frac{u_2}{u_1} = \frac{v_2}{v_1} = \frac{1 - z_1}{1 + z_1},$$

but $z_1 = \frac{u_1^2 + v_1^2 - 1}{1 + u_1^2 + v_1^2}$, therefore

$$\frac{u_2}{u_1} = \frac{v_2}{v_1} = \frac{1}{u_1^2 + v_1^2}.$$

Therefore

$$u_2 = \frac{u_1}{u_1^2 + v_1^2}, \quad v_2 = \frac{v_1}{u_1^2 + v_1^2}. \tag{5.3}$$

This implies that this mapping is a diffeomorphism. Thus, together these two charts provide an atlas for S^2.

We note that similar atlas, using stereographic mappings, can be introduced for the n-dimensional sphere S^n.

5.3.2 Grassmann Manifolds

These manifolds constitute a generalization of the projective plane.

Definition 5.9: A m-plane in R^n is a m-dimensional subspace of R^n which is determined by an ordered set of m-independent vectors in R^n

$$P_m = \{v = a_1 v_1 + \ldots + a_m v_m | v_i \in R^n, a_i \text{ constants}, i = 1, \ldots m\}.$$

For example, a plane through the origin in R^3 is a 2-plane in this space. (The plane has to contain the origin to be a subspace).

If P is $n \times m$ the matrix which has the vectors $\{v_1 \ldots v_m\}$ as column vectors then

$$P_m = \{\mathbf{x} \in R^n | \mathbf{x} = P\mathbf{a}, \mathbf{a} \in R^m\}.$$

From this representation it is clear that a matrix $Q = PA$ where A is a nonsingular $m \times m$ matrix will represent the same plan since A is a bijection $R^m \to R^m$.

We now define a manifold structure on the set $G(m, R^n)$ of all m-planes in R^n.

Consider all $P_m \in G(m, R^n)$ whose first m rows in the matrix P that represents P_m are independent. For each such P_m, we can find a nonsingular matrix so that PA is the form

$$\begin{pmatrix} I_m \\ Y \end{pmatrix},$$

where I_m is the Identity $m \times m$ matrix and $Y = [y_{ij}]$ is an $(n - m) \times m$ matrix. We refer to this representation as the canonical representation P_m. The mapping

$$Y \to \{y_{11} \ldots y_{(m-n)m} \in R^{(m-n)m}\},$$

is obviously a chart on the set of these elements of $G(m, R^n)$ (whose first m rows are linearly independent).

In the more general case where the rows $\{\alpha_1 < \ldots < \alpha_p\}$ in P are independent we can find a nonsingular matrix that brings these row into the unit matrix and the remaining rows will then form the matrix Y. A chart on this set of elements of $G(m, R^n)$ will map Y to $R^{(m-n)m}$ as before.

Obviously, the set of all these charts covers $G(m, R^n)$, and it is not hard to show that they actually form a C^∞ atlas.

Observe that the dimension of $G(m, R^n)$ is $m(n - m)$.

Example 5.12: Consider $G(2, R^3)$ the manifold of all two-dimensional planes through the origin in R^3. Any such plane can be represented by a 3×2 matrix P consisting of two vectors that generate the plane. Since these two generating vectors are not unique we can multiply P by any nonsingular 2×2 matrix to obtain the same plane. The canonical representation of a plane will belong to at least one of the following three sets: $\{U_{12}, U_{13}, U_{23}\}$

$$
\begin{pmatrix} 1 & 0 \\ 0 & 1 \\ x_{31} & x_{32} \end{pmatrix}, \quad
\begin{pmatrix} 1 & 0 \\ y_{21} & y_{22} \\ 0 & 1 \end{pmatrix}, \quad
\begin{pmatrix} z_{11} & z_{12} \\ 1 & 0 \\ 0 & 1 \end{pmatrix}.
$$

A chart on U_{12} is defined by a mapping X_{12} of the canonical representation of $P \in U_{12} \to (x_{31}, x_{32})$, etc. Hence, the three charts $X_{12}, X_{23}, X13$ on U_{12}, U_{23}, U_{13} respectively cover $G(2, R^3)$. To prove that they form a C^∞ atlas we have to show that on the intersection of these sets e.g on $U_{13} \cap U_{23}$ the mapping $X_{13} \circ X_{23}^{-1}$ is a diffeomorphism $R^2 \to R^2$. To show this consider $P \in U_{13} \cap U_{23}$. As a member of U_{23}, P has a canonical representation

$$
P = \begin{pmatrix} z_{11} & z_{12} \\ 1 & 0 \\ 0 & 1 \end{pmatrix}.
$$

However, since it is also a member of U_{13} we infer that $z_{11} \neq 0$. To obtain its canonical representation in U_{23}, we form the equivalent canonical representation by

$$
\begin{pmatrix} z_{11} & z_{12} \\ 1 & 0 \\ 0 & 1 \end{pmatrix}
\begin{pmatrix} z_{11} & z_{22} \\ 0 & 1 \end{pmatrix}^{-1}
= \begin{pmatrix} 1 & 0 \\ y_{21} & y_{22} \\ 0 & 1 \end{pmatrix},
$$

where $y_{23} = 1/z_{11}$, $y_{22} = -z_{12}/z_{11}$. This mapping is a C^∞ diffeomorphism since $z_{11} \neq 0$.

Theorem 5.3: For $0 < m < n$ the manifolds $G(m, R^n)$ and $G(n - m, R^n)$ are diffeomorphic.

Proof: We observe first that both manifolds have the same dimension.

Let $A \in G(m, R^n)$ and $B \in G(n - m, R^n)$. We shall say that A, B are orthogonal if $A^T B = \mathbf{0}$ where A^T is the transpose of A. Note that this relationship is independent of the representatives chosen for A, B. (Since if $A' = AP$

and $B' = BQ$ then $(A')^T B' = \mathbf{0}$.) It follows then that the solution space of $A^T \mathbf{z} = \mathbf{0}$ is spanned by the column vectors of B. The mapping $A \to B$ is a bijection, and it can be shown to be a diffeomorphism.

Definition 5.10: $P^n(R) = G(1, R^{n+1})$ is called the real projective space. It has dimension n and its structure is defined by an atlas of $(n+1)$ charts on $U_1, \ldots, U_{n+1}$

$$U_1 : (1, x_1, \ldots, x_n)^T, \quad \ldots, U_{n+1} : (z_1, \ldots, z_n, 1)^T.$$

Example 5.13: $P^2(R)$ ($=$ lines in R^3) admits atlas with domains U_1, U_2, U_3. A line $P \in U_1 \cap U_3$ has two equivalent representations

$$\begin{pmatrix} 1 \\ x_2 \\ x_3 \end{pmatrix}, \begin{pmatrix} z_1 \\ z_2 \\ 1 \end{pmatrix}.$$

The change of coordinates on $U_1 \cap U_3$ is given by

$$z_1 = 1/x_3, \quad z_2 = x_2/x_3.$$

(Observe that $x_3 \neq 0$).

Grassmann manifolds can be defined not only over R^n but also over C^n and Q^n (where Q is the quaternionic number system).

Theorem 5.4: $P^1(R)$ and $P^1(C)$ are diffeomorphic to the circle S^1 and the sphere S^2, respectively. To prove this theorem, one can use stereographic projections.

Exercises

1. Use the inclusion map to show that any open set S in R^n a manifold

2. Show that the set $M(n, n)$ of $n \times n$ matrices with real entries. is a manifold Hint: Consider the mapping

$$M(n, n) \to R^{n^2}.$$

6

Differentiation on Manifolds

6.1 Differentiation

Let M, N be (smooth) differential manifolds and f a mapping $f : M \to N$. If $\mathbf{x}$, $\mathbf{y}$ are charts on M, N, respectively, then a coordinate representation of this mapping on these charts is defined by

$$F : \mathbf{y} \circ f \circ \mathbf{x}^{-1} : R^n \to R^k.$$

When $N = R$ with the standard manifold structure (viz. $\mathbf{y}$ is the identity chart on R) we can rewrite this formula as $f = F \circ \mathbf{x}$. Using this we make the following definition

Definition 6.1: Let $f : M \to R$ with coordinate representative $F : R^n \to R$. We define

$$\frac{\partial f}{\partial x^i} = F_{,i} \circ \mathbf{x} : M \to R,$$

where $F_{,i} = \frac{\partial F}{\partial x^i}$.

Observe that this definition satisfies the usual properties of the derivative

$$\frac{\partial}{\partial x^i}(af + bg) = a\frac{\partial f}{\partial x^i} + b\frac{\partial g}{\partial x^i},$$

$$\frac{\partial}{\partial x^i}(fg) = \frac{\partial f}{\partial x^i}g + f\frac{\partial g}{\partial x^i},$$

where f, g are mappings $M \to R$ and a, b are constants.

Example 6.1: Let M be a manifold and $\mathbf{x}$ be a chart containing the point m viz. $\mathbf{x}(m) = (x^1(m), \dots, x^n(m))$.

Suppose that F is a function $F : R^n \to R$ which is defined as $F(x^1, \dots, x^n) = cos(x^1) + sin(x^2)$. The coordinate representation of $f = F \circ \mathbf{x} : M \to R$ at the point m is

$$f(m) = \cos(x^1(m)) + \sin(x^2(m)).$$

Hence

$$\frac{\partial f}{\partial x^1}(m) = -\sin(x^1(m)), \quad \frac{\partial f}{\partial x^2}(m) = \cos(x^2(m)).$$

DOI: 10.1201/9781003587422-6

6.2 Tangent Vectors

Let $\mathcal{F}(m)$ be the set of all differentiable functions $f : M \to R$ whose domain contain $m \in M$. We now consider linear operators that map this set to R.

Remark A linear operator L on $\mathcal{F}(m)$ have the following property

$$L(af + bg) = aL(f) + bL(g),$$

where a, b are constants.

Theorem 6.1: Let L be a linear operator on $\mathcal{F}(m)$ to R. If $f, g \in \mathcal{F}(m)$ agree on some neighborhood of m then $(Lf)(m) = (Lg)(m)$.

Definition 6.2: A derivation on $\mathcal{F}(m)$ is a linear operator $L : \mathcal{F}(m) \to R$ which satisfies

$$L(fg) = f(m)L(g) + L(f)g(m).$$

From the previous definition of the derivative of a function $f : M \to R$ we now see that

$$L = \frac{\partial}{\partial x^i} : \mathcal{F}(m) \to R,$$

which is defined by $Lf = \partial f / \partial x^i$ is a derivation on $\mathcal{F}(m)$.

If L_1, L_2 are two derivations on $\mathcal{F}(m)$ we define

$$aL_1 + bL_2 : \mathcal{F}(m) \to R,$$

by

$$(aL_1 + bL_2)(f) = aL_1(f) + bL_2(f).$$

With this structure the set of **all** derivations on $\mathcal{F}(m)$ is a vector space which we call the tangent space $T_m M$ at m. A derivation L in this space will be referred to as a "tangent vector at m".

Observe that the form of the general element in $T_m M$ is

$$L = \sum_{i=1}^{n} F_i \left(\frac{\partial}{\partial x^i} \right)_m.$$

where F_i are functions $R^n \to R$ The following elements of $T_m M$ $\{\frac{\partial}{\partial x^i}, i = 1, \ldots, n\}$ will be referred to as the canonical basis of $T_m M$ with respect to the chart x.

Observe that elements of the tangent space are now operators NOT "classical verctor". In many ways, this **NEW Definition** can be considered as the "bifurcation point" between classical and modern differential geometry.

Example 6.2: When $M = R^n$, we can write

$$L(f)(\mathbf{x}) = \mathbf{a} \cdot \mathrm{grad}(f)(\mathbf{x}).$$

Since $\frac{\partial x^j}{\partial x^i} = \delta_{ij}$ it follows that $L(x^j) = a_j$ and we can rewrite an operator L as

$$L = \sum_{i=1}^{n} L(x^i) \left(\frac{\partial}{\partial x^i} \right)_m. \tag{6.1}$$

If y is another chart at m this representation shows that

$$\frac{\partial}{\partial y^j} = \sum_{i=1}^{n} \left(\frac{\partial x^i}{\partial y^j} \right)_m \left(\frac{\partial}{\partial x^i} \right)_m.$$

Example 6.3: Compute the tangent plane to the one-sheet-hyperboloid M $x^2 + y^2 - z^2 = 1$.

Solution: Since the hyperboloid M is a two-dimensional manifold a (global) chart is a mapping $\mathbf{x} : M \to R^2$. Such a chart is given by

$$x = \cosh\phi \cos\theta, \quad y = \cosh\phi \sin\theta, \quad z = \sinh\phi.$$

The tagent plane to M at a point m is generated by

$$X = \left(\frac{\partial}{\partial\phi} \right)_m, \quad Y = \left(\frac{\partial}{\partial\theta} \right)_m,$$

and the general vector in TM_m is $(aX + bY)_m$. To compute the representation of this vector in cartesian coordinate we observe that

$$\frac{\partial}{\partial\phi} = \sinh\phi \cos\theta \frac{\partial}{\partial x} + \sinh\phi \sin\theta \frac{\partial}{\partial y} + \cosh\phi \frac{\partial}{\partial z},$$

$$\frac{\partial}{\partial\theta} = -\cosh\phi \sin\theta \frac{\partial}{\partial x} + \cosh\phi \cos\theta \frac{\partial}{\partial y}.$$

Hence

$$X = \frac{xz}{r} \frac{\partial}{\partial x} + \frac{yz}{r} \frac{\partial}{\partial y} + r\frac{\partial}{\partial z}, \quad r = \sqrt{x^2 + y^2},$$

$$Y = x\frac{\partial}{\partial y} - y\frac{\partial}{\partial x}.$$

Example 6.4: Let $M = S^2$. Use the sterographic charts on S^2 to compute the tangent plane on this manifold and derive the explicit expressions for X_1, Y_1 in terms of X_2, Y_2 (see Chapter 5)

Solution: Using the sterographic chart on S^2 without the point $N = (0, 0, 1)$. ("pinched sphere" without the north pole. See chapter 5) we have

$$x_1 = u_1 t_1, \quad y_1 = v_1 t_1, z_1 = 1 - t_1,$$

where $t_1 = \frac{2}{1+u_1^2+v_1^2}$. Hence

The tangent plane on this chart at a point $m = (x_1, y_1, z_1)$ is generated by

$$X_1 = \left(\frac{\partial}{\partial u_1}\right)_m, \quad Y_1 = \left(\frac{\partial}{\partial v_1}\right)_m.$$

Similarly on the sterographic chart on S^2 without the point $M = (0,0,-1)$ we have

$$X_2 = \left(\frac{\partial}{\partial u_2}\right)_m, \quad Y_2 = \left(\frac{\partial}{\partial v_2}\right)_m.$$

On the intersection of these two charts, we can re-express X_1 and Y_1 in terms of X_2, Y_2,

$$\left(\frac{\partial}{\partial u_1}\right)_m = \left(\frac{\partial u_2}{\partial u_1}\right)_m \left(\frac{\partial}{\partial u_2}\right)_m + \left(\frac{\partial v_2}{\partial u_1}\right)_m \left(\frac{\partial}{\partial v_2}\right)_m,$$

and

$$\left(\frac{\partial}{\partial v_1}\right)_m = \left(\frac{\partial u_2}{\partial v_1}\right)_m \left(\frac{\partial}{\partial u_2}\right)_m + \left(\frac{\partial v_2}{\partial v_1}\right)_m \left(\frac{\partial}{\partial v_2}\right)_m.$$

Using Eq. (5.3) in Chapter 5, we have

$$\left(\frac{\partial u_2}{\partial u_1}\right)_m = \frac{-u_1^2 + v_1^2}{(u_1^2 + v_1^2)^2}, \quad \left(\frac{\partial v_2}{\partial u_1}\right)_m = \frac{-2v_1 u_1}{(u_1^2 + v_1^2)^2},$$

$$\left(\frac{\partial u_2}{\partial v_1}\right)_m = \frac{-2v_1 u_1}{(u_1^2 + v_1^2)^2}, \quad \left(\frac{\partial v_2}{\partial v_1}\right)_m = \frac{u_1^2 - v_1^2}{(u_1^2 + v_1^2)^2}.$$

Hence

$$X_1 = \frac{1}{(u_1^2 + v_1^2)^2}[(-u_1^2 + v_1^2)X_2 - (2v_1 u_1)Y_2]$$

$$Y_1 = \frac{1}{(u_1^2 + v_1^2)^2}[-(2v_1 u_1)X_2 + (u_1^2 + v_1^2)^2 Y_2].$$

6.3 Derived Mappings

Let ψ be a mapping $M \to N$ and let $\psi(m) = n \in N$. We now want to define a "lift" this mapping ψ to a mapping between the tangent spaces at m and n viz.$\psi_* : T_m M \to T_n N$.

Let $f' \in \mathcal{F}(n)$. The function $f = f' \circ \psi$ is in $\mathcal{F}(m)$. Therefore if $v \in T_m M$ then $v(f' \circ \psi)$ is well defined. We now claim that the mapping

$$w : f' \to v(f' \circ \psi),$$

is a derivation on $\mathcal{F}(n)$. To show this we note that if $f', g' \in \mathcal{F}(n)$, then

$$w(af' + bg') = v[(af' + bg') \circ \psi] = av(f' \circ \psi) + bv(g' \circ \psi),$$

$$w(f'g') = v(f'g' \circ \psi) = v[(f' \circ \psi)(g' \circ \psi)],$$

hence $w \in T_n N$. The mapping $\psi_* : v \to w$ is called the derived mapping of the mapping ψ or the differential of ψ.

We can summarize this definition by

$$\psi_*(v)(f') = v(f' \circ \psi).$$

To gain a concrete representation of this mapping let x, y be charts on M, N at m, n respectively. If $w = \psi_*(v)$ then from (6.1) we have the following coordinate representation for w

$$w = \sum_j (w(y^j)) \left(\frac{\partial}{\partial y^j} \right)_n,$$

but by definition

$$w(y^j) = \psi_*(v)(y^j) = v(y^j \circ \psi).$$

Hence the linear operator w that v is mapped to is

$$w = \sum_j (v(y^j \circ \psi)) \left(\frac{\partial}{\partial y^j} \right)_n,$$

When $v = \left(\frac{\partial}{\partial x^i} \right)_m$ is one of the canonical basis vectors of $T_m M$ we have then

$$v \to w = \sum_j \frac{\partial(y^j \circ \psi)}{\partial x^i} \left(\frac{\partial}{\partial y^j} \right)_n.$$

Observe that in this formula ψ stands for the composition of $\psi \circ \mathbf{x}$ where $\mathbf{x}$ represents the chart coordinates.

We note that the matrix

$$J(n) = \left[\frac{\partial(y^j \circ \psi)}{\partial x^i} \right],$$

is the Jacobian matrix representation of $\Psi = y \circ \psi \circ x^{-1}$.

Example 6.5: let $\psi : R^2 \to R$ be defined as

$$(x_1, x_2) \to 2x_1^3 - x_1 x_2 - x_2^3 + 1,$$

compute the map $\psi_* : T_m R^2 \to T_{\phi(m)} R$

Solution: The tangent space of R has only one dimension and its canonical basis is $\frac{\partial}{\partial t}$. Moreover, since R is one-dimensional $y^1 \circ \psi = \psi$. Hence

$$\psi_* \left(\frac{\partial}{\partial x_1} \right)_p = (6x_1^2 - x_2)(p) \left(\frac{\partial}{\partial t} \right)_{\psi(p)},$$

$$\psi_* \left(\frac{\partial}{\partial x_2} \right)_p = (-x_1 - 3x_2^2)(p) \left(\frac{\partial}{\partial t} \right)_{\psi(p)},$$

When $p = (1,1)$ this yields

$$\psi_* \left(\frac{\partial}{\partial x_1} \right)_{(1,1)} = 5 \left(\frac{\partial}{\partial t} \right)_1,$$

$$\psi_* \left(\frac{\partial}{\partial x_2} \right)_{(1,1)} = -4 \left(\frac{\partial}{\partial t} \right)_1.$$

Example 6.6: Let $\psi : R^2 \to R^3$ be defined by

$$(x_1, x_2) \to (x_1^2 x_2^2 + x_2^3, x_1 - 2x_2, x_1^2 e^{x_2}),$$

compute ψ_*

Solution: The canonical basis for $T_{\psi(p)} R^3$ is given by $\left\{ \frac{\partial}{\partial y_1}, \frac{\partial}{\partial y_2}, \frac{\partial}{\partial y_3} \right\}$. Hence,

$$\psi_* \left(\frac{\partial}{\partial x_i} \right)_m = \sum_j \frac{\partial(y_j \circ \psi)}{\partial x_i} \left(\frac{\partial}{\partial y_j} \right)_{\psi(m)}.$$

Since $(y_1 \circ \psi) = x_1^2 x_2^2 + x_2^3$, $(y_2 \circ \psi) = x_1 - 2x_2$, $(y_3 \circ \psi) = x_1^2 e^{x_2}$ we have

$$\psi_* \left(\frac{\partial}{\partial x_1} \right)_p = (2x_1 x_2^2) \left(\frac{\partial}{\partial y_1} \right)_{\psi(p)} + \left(\frac{\partial}{\partial y_2} \right)_{\psi(p)} + 2x_1 e^{x_2} \left(\frac{\partial}{\partial y_3} \right)_{\psi(p)},$$

$$\psi_* \left(\frac{\partial}{\partial x_2} \right)_p = (2x_1^2 x_2 + 3x_2^2) \left(\frac{\partial}{\partial y_1} \right)_{\psi(p)} - 2 \left(\frac{\partial}{\partial y_2} \right)_{\psi(p)} + x_1^2 e^{x_2} \left(\frac{\partial}{\partial y_3} \right)_{\psi(p)}.$$

In particular, the image of

$$\psi_* \left[2\left(\frac{\partial}{\partial x_1} \right) - 3 \left(\frac{\partial}{\partial x_2} \right) \right]_p,$$

at $p = (1,0)$ is

$$\left[8 \frac{\partial}{\partial y_2} + \frac{\partial}{\partial y_3} \right]_{(0,1,1)}.$$

Example 6.7: A vector $(x_1, x_2, x_3, x_4) \in R^4$ can be written in matrix form as

$$\mathbf{X} = \begin{pmatrix} x_1 & x_2 \\ x_3 & x_4 \end{pmatrix}.$$

Consider the following transformation $\psi : R^4 \to R^4$ which is defined as $\mathbf{X} \to \mathbf{TX}$ where

$$\mathbf{T} = \begin{pmatrix} \cos\theta & \sin\theta \\ -\sin\theta & \cos\theta \end{pmatrix}.$$

Compute $\mathbf{T}_*$

Solution: The explicit form of the transformation T is

$$\mathbf{TX} \to (x_1\cos\theta + x_3\sin\theta, \, x_2\cos\theta + x_4\sin\theta, \, -x_1\sin\theta + x_3\cos\theta, \, -x_2\sin\theta + x_4\cos\theta).$$

Hence

$$\mathbf{T}_*\left(\frac{\partial}{\partial x_1}\right)_p = \cos\theta\left(\frac{\partial}{\partial y_1}\right)_{\psi(p)} - \sin\theta\left(\frac{\partial}{\partial y_3}\right)_{\psi(p)},$$

$$\mathbf{T}_*\left(\frac{\partial}{\partial x_2}\right)_p = \cos\theta\left(\frac{\partial}{\partial y_2}\right)_{\psi(p)} - \sin\theta\left(\frac{\partial}{\partial y_4}\right)_{\psi(p)},$$

$$\mathbf{T}_*\left(\frac{\partial}{\partial x_3}\right)_p = \sin\theta\left(\frac{\partial}{\partial y_1}\right)_{\psi(p)} + \cos\theta\left(\frac{\partial}{\partial y_3}\right)_{\psi(p)},$$

$$\mathbf{T}_*\left(\frac{\partial}{\partial x_4}\right)_p = \sin\theta\left(\frac{\partial}{\partial y_2}\right)_{\psi(p)} + \cos\theta\left(\frac{\partial}{\partial y_4}\right)_{\psi(p)}.$$

Example 6.8: Let $f : R^3 \to R$ be defined as $f(x,y,z) = x^3 y$ and let $X \in TR^3$ be defined as

$$X = xy\left(\frac{\partial}{\partial x}\right) + x^3\left(\frac{\partial}{\partial z}\right).$$

Compute $f_* X_{(1,1,0)}$

Solution: The tangent space on R is one-dimensional and it is generated by $\frac{\partial}{\partial t}$. On the other hand, the tangent space on R^3 has a canonical basis $\left\{\frac{\partial}{\partial x}, \frac{\partial}{\partial y}, \frac{\partial}{\partial z}\right\}$. Hence

$$f_*\left(\frac{\partial}{\partial x}\right)_{\mathbf{x}} = \left(\frac{\partial f}{\partial x}\right)_{\mathbf{x}}\left(\frac{\partial}{\partial t}\right)_{f(\mathbf{x})} = (3x^2 y)_{\mathbf{x}}\left(\frac{\partial}{\partial t}\right)_{f(\mathbf{x})}.$$

Similarly, we have

$$f_*\left(\frac{\partial}{\partial y}\right)_{\mathbf{x}} = (x^3)_{\mathbf{x}}\left(\frac{\partial}{\partial t}\right)_{f(\mathbf{x})},$$

and

$$f_*\left(\frac{\partial}{\partial z}\right)_{\mathbf{x}} = 0.$$

Therefore,

$$f_* X_{(1,1,0)} = 3\left(\frac{\partial}{\partial t}\right)_{(1)}.$$

7

Vectors and Bundles

7.1 The Tangent Bundle

Let M be a differential manifold. The union of all tangent spaces $T_m M$ for all $m \in M$ is called the **Tangent Bundle** TM.

$$TM = \{(m, X_m) | m \in M, X_m \in T_m M\}.$$

The projection

$$\pi : TM \to M,$$

is defined as $\pi(m, X_m) = m$.

A mapping

$$\psi : M \to N,$$

induces a mapping

$$\psi_* : TM \to TN,$$

which is defined by $(m, X_m) \to (\psi(m), \psi_*(X_m))$ The mapping ψ_* is referred to as the differential of ψ.

We now show how a C^∞ structure can be defined on TM i.e TM is a manifold.

Let x be a chart on M with domain U. A vector field $X \in \pi^{-1}U$ can be expressed uniquely as

$$X_m = \sum a_i \left(\frac{\partial}{\partial x_i} \right)_m, \quad \mathbf{a} = (a_1, \ldots, a_n) \in R^n.$$

We have therefore an injection

$$\psi : TM \to R^{2n},$$

which maps $(m, X_m) \to (x(m), \mathbf{a})$. The domain of this map is $\pi^{-1}U$, and its range is an open set $x(U) \times R^n$. These "standard" charts define a C^∞ atlas on TM. Therefore, TM is manifold of dimension $2n$.

DOI: 10.1201/9781003587422-7

Theorem 7.1: If $\phi : M \to N$ is differentiable then $\phi_* : TM \to TN$ is also differentiable.

As explained earlier, the explicit representation of ϕ_* is

$$\phi_*(m, \mathbf{v}) = (\phi(m), \mathbf{w}),$$

where for any $f' \in \mathcal{F}(\phi(m))$ $\mathbf{w}$ satisfies $\mathbf{w}(f')(\phi(m)) = v(f' \circ \phi)(m)$

7.2   Cotangent Bundle

7.2.1   Brief Linear Algebra Review

Let $V = R^n$. A linear functional on V is a mapping

$$f : V \to R,$$

so that

$$f(a\mathbf{v} + b\mathbf{w}) = af(\mathbf{v}) + bf(\mathbf{w}), \quad a, b \in R.$$

The set of all linear functionals on V is a vector space with addition and multiplication by scalar given by

$$(af)(\mathbf{v}) = f(a\mathbf{v}), \quad (af + bg)(\mathbf{v}) = af(\mathbf{v}) + bg(\mathbf{v}).$$

This vector space is called the dual of V and is denoted by V^*. The dimension of V^* is equal to the dimension of V. If $\{\mathbf{v}_1, \dots, \mathbf{v}_n\}$ is a basis of V the dual basis of V^* is defined by

$$\omega_i(\mathbf{v}_j) = \delta_{ij}, \quad i, j = 1, \dots, n.$$

We observe also that $(V^*)^* = V$

7.2.2   Cotangent Bundle on a Manifold

Let M be a manifold. At each point $m \in M$, we defined the tangent space TM_m which is isomorphic to R^n. The dual space to TM_m is denoted by TM_m^*. The **cotangent bundle** on M is defined as

$$TM^* = \cup_m TM_m^*,$$

i.e. the elements of TM^* are $\tau = (m, \omega)$ where $m \in M$ and $\omega \in TM_m^*$ with the natural projection

$$\pi : TM^* \to M, \quad \pi(\tau) = \pi(m, \omega) = m.$$

A differentiable structure on TM^* can be defined as follows: Let $\{U, \phi\}$ be a chart on M with coordinate functions $x_1, \ldots, x_n$. We define the corresponding chart on TM^* by

$$\phi^* : (U, \cup_m TM_m^*) \to R^{2n}, \quad m \in U,$$

$$\phi^*(\tau) = \left(x_1(m), \ldots, x_n(m), \omega \left(\frac{\partial}{\partial x_1} \right), \ldots, \omega \left(\frac{\partial}{\partial x_n} \right) \right).$$

In this context the dual basis to $\frac{\partial}{\partial x_i}$, $i = 1, \ldots, n$ is denoted by dx_i i.e.

$$dx_i \left(\frac{\partial}{\partial x_j} \right) = \delta_{ij}.$$

If $\phi : M \to N$ we can lift this mapping to the corresponding cotangent bundles and define

$$\phi^* : TN^* \to TM^*,$$

as follows

$$\phi^*(\omega)(X) = \phi_*(X)(\omega), \quad \omega \in TN^*, \quad X \in TM.$$

Note that $\phi_*(X) \in TN = (TN^*)^*$. Observe also the reverse direction of this mapping as compared to ϕ_*

We can have now the following diagram:

$$\phi^* : TM^* \leftarrow TN^*$$
$$\uparrow \qquad \uparrow$$
$$\phi : M \quad \to \quad N$$
$$\downarrow \qquad \downarrow$$
$$\phi_* : TM \to TN.$$

Example 7.1: Let

$$\omega = (2xy + x + 1)dx + (x - y)dy \in (R^2)^*,$$

and

$$\phi : R^3 \to R^2, \quad (u, v, w) \to (x, y) = (u + v, v - w).$$

Compute $\phi^*(\omega)$

Solution: A direct (calculus type) approach will be to use the definition of the map ϕ to derive $dx = du + dv$, $dy = dv - dw$. Substituting this in the definition of ω and replacing x, y by u, v, w we obtain

$$\begin{aligned} \phi^*(\omega) &= (2(u+v)(v-w) + u + v + 1)du + (2(u+v)(v-w) \\ &\quad + 2u + v + 1 + w)dv - (u+w)dw. \end{aligned}$$

A formal approach to the solution of this problem will be as follows:

We compute first

$$\phi^*(dx) = a_1 du + b_1 dv + c_1 dw, \quad \phi^*(dy) = a_2 du + b_2 dv + c_2 dw,$$

where

$$a_1 = \phi^*(dx)\left(\frac{\partial}{\partial u}\right), \quad b_1 = \phi^*(dx)\left(\frac{\partial}{\partial v}\right), \quad c_1 = \phi^*(dx)\left(\frac{\partial}{\partial w}\right).$$

with similar expressions for a_2, b_2, c_2.

$$
\begin{aligned}
a_1 &= \phi^*(dx)\left(\frac{\partial}{\partial u}\right) = \phi_*\left(\frac{\partial}{\partial u}\right)(dx) \\
&= \left[\frac{\partial(x \circ \phi)}{\partial u}\frac{\partial}{\partial x} + \frac{\partial(y \circ \phi)}{\partial u}\frac{\partial}{\partial y}\right](dx) = \left(\frac{\partial}{\partial x}(dx)\right) = 1.
\end{aligned}
$$

Similarly

$$
\begin{aligned}
b_1 &= \phi^*(dx)\left(\frac{\partial}{\partial v}\right) = \phi_*\left(\frac{\partial}{\partial v}\right)(dx) \\
&= \left[\frac{\partial(x \circ \phi)}{\partial v}\frac{\partial}{\partial x} + \frac{\partial(y \circ \phi)}{\partial v}\frac{\partial}{\partial y}\right](dx) = \left(\frac{\partial}{\partial x} + \frac{\partial}{\partial y}\right)(dx) = 1.
\end{aligned}
$$

Finally

$$
\begin{aligned}
c_1 &= \phi^*(dx)\left(\frac{\partial}{\partial w}\right) = \phi_*\left(\frac{\partial}{\partial w}\right)(dx) \\
&= \left[\frac{\partial(x \circ \phi)}{\partial w}\frac{\partial}{\partial x} + \frac{\partial(y \circ \phi)}{\partial w}\frac{\partial}{\partial y}\right](dx) = -\left(\frac{\partial}{\partial y}\right)(dx) = 0.
\end{aligned}
$$

Hence $\phi^*(dx) = du + dv$ etc.

Example 7.2: Let $\phi : R^2 \to R^2$ be defined as

$$(u, v) \to (x, y) = (uv, 1).$$

Compute $\phi^*(dx)$

Solution: We know that

$$\phi^*(dx) = adu + bdv.$$

Therefore

$$
\begin{aligned}
a &= \phi^*(dx)\left(\frac{\partial}{\partial u}\right) = \phi_*\left(\frac{\partial}{\partial u}\right)(dx) \\
&= \left[\frac{\partial(x \circ \phi)}{\partial u}\frac{\partial}{\partial x} + \frac{\partial(y \circ \phi)}{\partial u}\frac{\partial}{\partial y}\right](dx) = v\left(\frac{\partial}{\partial x}\right)(dx) = v.
\end{aligned}
$$

Similarly

$$
\begin{aligned}
b &= \phi^*(dx)\left(\frac{\partial}{\partial v}\right) = \phi_*\left(\frac{\partial}{\partial v}\right)(dx) \\
&= \left[\frac{\partial(x \circ \phi)}{\partial v}\frac{\partial}{\partial x} + \frac{\partial(y \circ \phi)}{\partial v}\frac{\partial}{\partial y}\right](dx) = u\left(\frac{\partial}{\partial x}\right)(dx) = u.
\end{aligned}
$$

Hence

$$
\phi^*(dx) = vdu + udv.
$$

Example 7.3: For the same transformation given in the previous example compute $\phi^*(ydx)$

Solution: As before we have

$$
\begin{aligned}
\phi^*(dx) &= adu + bdv, \\
a &= \phi^*(ydx)\left(\frac{\partial}{\partial u}\right) = \phi_*\left(\frac{\partial}{\partial u}\right)(ydx) \\
&= \left[\frac{\partial(x \circ \phi)}{\partial u}\frac{\partial}{\partial x} + \frac{\partial(y \circ \phi)}{\partial u}\frac{\partial}{\partial y}\right](ydx) = v\left(\frac{\partial}{\partial x}\right)(ydx) = v \star 1.
\end{aligned}
$$

Similarly

$$
\begin{aligned}
b &= \phi^*(dx)\left(\frac{\partial}{\partial v}\right) = \phi_*\left(\frac{\partial}{\partial v}\right)(ydx) \\
&= \left[\frac{\partial(x \circ \phi)}{\partial v}\frac{\partial}{\partial x} + \frac{\partial(y \circ \phi)}{\partial v}\frac{\partial}{\partial y}\right](ydx) = u\left(\frac{\partial}{\partial x}\right)(ydx) = u \star 1.
\end{aligned}
$$

Hence

$$
\phi^*(ydx) = vdu + udv.
$$

Example 7.4: Let $\omega = zdx$ and let $\psi : R^2 \to R^3$ be defined as

$$
\psi(\phi, \theta) \to (x, y, z) = (\sin\phi\cos\theta, \sin\phi\sin\theta, \cos\phi),
$$

compute $\psi^*(\omega)$

Solution: We know that

$$
\psi^*(\omega) = ad\phi + bd\theta.
$$

We compute a, b

$$
\begin{aligned}
a &= \psi^*(\omega)\left(\frac{\partial}{\partial\phi}\right) = \psi_*\left(\frac{\partial}{\partial\phi}\right)(zdx) \\
&= \left[\frac{\partial(x \circ \psi)}{\partial\phi}\frac{\partial}{\partial x} + \frac{\partial(y \circ \psi)}{\partial\phi}\frac{\partial}{\partial y} + \frac{\partial(z \circ \psi)}{\partial\phi}\frac{\partial}{\partial z}\right](zdx) \\
&= \cos\phi\cos\theta\left(\frac{\partial}{\partial x}\right)(zdx) = \cos^2\phi\cos\theta,
\end{aligned}
$$

$$b = \psi^*(\omega)\left(\frac{\partial}{\partial\theta}\right) = \psi_*\left(\frac{\partial}{\partial\theta}\right)(zdx)$$

$$= \left[\frac{\partial(x\circ\psi)}{\partial\theta}\frac{\partial}{\partial x} + \frac{\partial(y\circ\psi)}{\partial\theta}\frac{\partial}{\partial y} + \frac{\partial(z\circ\psi)}{\partial\theta}\frac{\partial}{\partial z}\right](zdx)$$

$$= -\sin\phi\sin\theta\left(\frac{\partial}{\partial x}\right)(zdx) = -\sin\phi\cos\phi\cos\theta.$$

Hence

$$\psi^*(\omega) = \cos^2\phi\cos\theta d\phi - \sin\phi\cos\phi\cos\theta d\theta.$$

7.3   Vector Fields

Definition 7.1: A section X of TM is a mapping $M \to TM$, which associates with every point of its domain a vector $X(m) \in T_mM$.

Let $f : M \to R$ be differentiable. We define $Xf : M \to R$ on points m which are in the intersections of the domains of X and f as $m \to X(m)f$

Definition 7.2: A section X is called a vector field on M if Xf are differentiable for all $f : M \to R$ whose domain intersects the domain of X

Example 7.5: On the domain of a chart x of M the mapping $m \to X(m) = (\frac{\partial}{\partial x_i})_m$ is a vector field since by definition

$$Xf = \frac{\partial f}{\partial x_i},$$

which is differentiable.

Example 7.6: Let X, Y be vector fields on M and $f, g : M \to R$ differentiable then (on the intersection of the domains) $fX + gY$ which is defined by $m \to f(m)X(m) + g(m)Y(m)$ is a section of TM. It is a vector field if f, g are differentiable.

To make this even more concrete, we observe that any tangent vector in T_mM can be expressed uniquely as

$$X(m) = \sum X(x^i)_m\left(\frac{\partial}{\partial x_i}\right)_m.$$

If X is a section of TM then X is a vector field if the functions $X(x^i)$ are differentiable on the chart x.

If we want to express the vector field in terms of another chart (whose domain intersects the original chart), we have the chain rule

$$\frac{\partial}{\partial y_i} = \sum \frac{\partial x_j}{\partial y_i} \frac{\partial}{\partial x_j}.$$

Definition 7.3: Let $\mathcal{F}_U$ be the set of all differentiable functions on an open set $U \subset M$. A differential operator L on $\mathcal{F}_U$ is a linear operator on $\mathcal{F}_U \to \mathcal{F}_U$ which satisfies

$$L(fg) = fL(g) + gL(f),$$

for all $f, g \in \mathcal{F}_U$

It is obvious that any vector field X on U defines a differential operator on $\mathcal{F}_U$ by $X : f \to Xf$ since Xf is differentiable. We also have the converse

Theorem 7.2: Any differential operator on U corresponds to a unique vector field on U.

Let X, Y be vector fields on $U \subset M$. We now show that

$$Lf = X(Yf) - Y(Xf),$$

is a differential operator on $\mathcal{F}_U$.

In fact;

$$\begin{aligned} L(fg) =& X(Y(fg)) - Y(X(fg)) = X(gY(f) + fY(g)) - Y(gX(f) + fX(g)) \\ =& X(g)Y(f) + gX(Y(f)) + X(f)Y(g) + fX(Y(g)) \\ & -Y(g)X(f) - gY(X(f)) - Y(f)X(g) - fY(X(g)). \end{aligned}$$

Hence

$$L(fg) = f[X(Y(g)) - Y(X(g))] + g[X(Y(f)) - Y(X(f))] = fL(g) + gL(f).$$

It is obvious also that

$$L(af + bg) = aL(f) + bL(g), \quad a, b \in R.$$

We conclude therefore that L is a linear operator on $U \subset M$ and therefore there is a unique vector field that correspond to it. We denote this vector field by $[X, Y]$. Observe that the mapping $(X, Y) \to [X, Y]$ is bilinear (i.e., linear in each component).

We showed therefore that the set of all linear operators on $\mathcal{F}_U$ is a vector space, and it admits a skew symmetric multiplication $(X, Y) \to [X, Y]$. Another important property of this multiplication is the **Jacobi Identity**

$$[X, [Y, Z]] + [Y, [Z, X]] + [Z, [X, Y]] = 0.$$

Example 7.7: Show that on a chart ϕ whose domain is U

$$L = \left[\frac{\partial}{\partial x_i}, \frac{\partial}{\partial x_j}\right] = 0.$$

Solution: It is enough to show that for any $p \in U$ and $f \in \mathcal{F}_U$ we have $L_p(f) = 0$.

By definition $f = F(\mathbf{x}) \circ \phi$ and

$$\frac{\partial f}{\partial x_i} = \frac{\partial F}{\partial x_i} \circ \phi,$$

therefore

$$\frac{\partial}{\partial x_i}\left(\frac{\partial f}{\partial x_j}\right) = \frac{\partial^2 F}{\partial x_i x_j} \circ \phi,$$

and the statement follows from the independence of the order of the partial derivatives.

Example 7.8: Find the general expression of the vector field X on $\mathcal{F}(R^2)$ which satisfies

$$\left[\frac{\partial}{\partial x}, X\right] = X.$$

Solution: The general form of X is

$$X = a(x,y)\frac{\partial}{\partial x} + b(x,y)\frac{\partial}{\partial y}.$$

Substituting this in the desired relationship for X we obtain

$$\frac{\partial a(x,y)}{\partial x} = a(x,y), \quad \frac{\partial b(x,y)}{\partial x} = b(x,y),$$

$$\frac{\partial a(x,y)}{\partial y} = a(x,y), \quad \frac{\partial b(x,y)}{\partial y} = b(x,y).$$

The first set of equations yields

$$a(x,y) = Af(y)e^x, \quad b(x,y) = Bg(y)e^x.$$

Substituting this in the second set of equations, we finally obtain

$$X = e^{x+y}\left[A\frac{\partial}{\partial x} + B\frac{\partial}{\partial y}\right],$$

where A, B are arbitrary constants.

Example 7.9: Let X, Y be vector fields on M. Compute [fX,gY] where $f, g \in \mathcal{F}_U$

Solution:

$$\begin{aligned}
[fX, gY](h) &= (fX)(gY(h)) - (gY)(fX(h)) \\
&= fX(g)Y(h) + fgX(Y(h)) - gY(f)X(h) - gfY(X(h)) \\
&= fX(g)Y(h) - gY(f)X(h) + fg[X, Y](h). \quad (7.1)
\end{aligned}$$

Hence

$$[fX, gY] = fX(g)Y - gY(f)X + fg[X, Y].$$

Example 7.10:

Consider the following vector fields on R^2

$$A = (x^2 + y)\frac{\partial}{\partial x} + (y^2 + 1)\frac{\partial}{\partial y}, \quad B = (y - 1)\frac{\partial}{\partial x},$$

and let μ be the following differential form on R^2

$$\mu = (2xy + x^2 + 1)dx + (x^2 - y)dy.$$

Furthermore let $\psi : R^3 \to R^2$ be defined as

$$\psi : (u, v, w) \to (x, y) = (u - v, v^2 + w).$$

Evaluate

$$a. \ [A, B]_{0,0}, \quad b. \ \mu(A)(0,0), \quad c. \ \psi^*(\mu).$$

Solution:

$$a. \quad [A, B] = [2x(1 - y) + y^2 + 1]\frac{\partial}{\partial x}.$$

Hence $[A, B]_{(0,0)} = \frac{\partial}{\partial x}|_{(0,0)}$.

$$\begin{aligned}
b. \ \mu(A) &= (2xy + x^2 + 1)dx(A) + (x^2 - y)dy(A) \quad (7.2) \\
&= [(2xy + x^2 + 1)(x^2 + y) + (y^2 + 1)(x^2 - y)]_{(0,0)} = 0.
\end{aligned}$$

$$\begin{aligned}
c. \ \psi^*\mu &= [2(u - v)(v^2 + w) + (u - v)^2 + 1]du \quad (7.3) \\
&+ 2v[(u - v)^2 - v^2 - w] - 2(u - v)(v^2 + w) - (u - v)^2 - 1dv \\
&+ [(u - v)^2 - v^2 - w]dw.
\end{aligned}$$

Example 7.11: Let A,B be vector fields on R^2

$$A = x\frac{\partial}{\partial x} + 2xy\frac{\partial}{\partial y}, \quad B = y\frac{\partial}{\partial y},$$

and let μ be the differential form

$$\mu = (x^2 + 2y)dx + (x + y^2)dy.$$

Compute

$$\mu(A, B) = A\mu(B) - B\mu(A) - \mu([A, B]).$$

Solution: We have

$$A\mu(B) = A(xy + y^3) = xy + 2x^2y + 6xy^3, \qquad (7.4)$$
$$B\mu(A) = B[x(x^2 + 2y) + 2xy(x + y^2)] = 2xy + 2x^2y + 6xy^3.$$

Also $\mu([A, B]) = \mu(0) = 0$ and therefore

$$\mu(A, B) = -xy.$$

8

Differential Forms

8.1 Exterior Products

The following can be done over a general manifold, but for simplicity, we restrict ourselves to a vector space V of dimension n.

Definition 8.1: Let $\mathbf{v}$, $\mathbf{w}$ be two vectors in V. We define the exterior (or wedge) product of these two vectors as (an abstract binary operation) $v \wedge w$ with the following properties:

1. $\mathbf{v} \wedge \mathbf{v} = 0$

2. $\mathbf{v} \wedge \mathbf{w} = -\mathbf{w} \wedge \mathbf{v}$

3. $(a\mathbf{v}_1 + b\mathbf{v}_2) \wedge \mathbf{w} = a(\mathbf{v}_1 \wedge \mathbf{w}) + b(\mathbf{v}_2 \wedge \mathbf{w}), \quad a, b \in R$

From these definitions it is obvious that if $\mathbf{w} = a\mathbf{v}$ then $\mathbf{v} \wedge \mathbf{w} = a(\mathbf{v} \wedge \mathbf{v}) = 0$.

Let $\{\mathbf{v}_1, \ldots, \mathbf{v}_n\}$ be a basis of V then any vector in V can be written as

$$\mathbf{v} = \sum_i a_i \mathbf{v}_i,$$

hence for any two vectors $\mathbf{v}, \mathbf{w}$ we have

$$\mathbf{v} \wedge \mathbf{w} = \left(\sum_i a_i \mathbf{v}_i \right) \wedge \left(\sum_j b_j \mathbf{v}_j \right) = \sum_{ij} a_i b_j \mathbf{v}_i \wedge \mathbf{v}_j.$$

Using the first two properties of the wedge product this can be rewritten as

$$\mathbf{v} \wedge \mathbf{w} = \sum_{i<j} (a_i b_j - a_j b_i)(\mathbf{v}_i \wedge \mathbf{v}_j).$$

We shall denote the set of wedge products of two vectors in V by $\bigwedge^2 V$. It is obviously a vector space and its basis is given by $\{\mathbf{v}_i \wedge \mathbf{v}_j, \ i < j\}$ i.e its dimension is $\frac{n(n-1)}{2}$.

DOI: 10.1201/9781003587422-8

Is there a relationship between wedge product and tensors? The answer is yes. If we look on a (covariant or contravariant) second order tensor

$$T_{ij} = \mathbf{v}_i \mathbf{v}_j,$$

and look at its antisymmetric part

$$A_{ij} = T_{ij} - T_{ji} = \mathbf{v}_i \mathbf{v}_j - \mathbf{v}_j \mathbf{v}_i,$$

then it is easy to see that A_{ij} satisfies all the properties of the wedge product.

In a similar manner we can define the wedge product of p vectors $\mathbf{w}_1 \ldots \mathbf{w}_p$ as (an abstract operation) $\mathbf{w}_1 \wedge \ldots, \wedge \mathbf{w}_p$ which has the following properties:

1. $\mathbf{w}_1 \wedge \ldots \wedge \mathbf{w}_p = 0$ if some indices $i \neq j$, $\mathbf{w}_i = \mathbf{w}_j$.

2. $\mathbf{w}_1 \wedge \ldots \wedge \mathbf{w}_p$ changes sign if the position of two vectors in the wedge product is interchanged.

3. $(a\mathbf{u} + b\mathbf{w}) \wedge \mathbf{w}_2 \wedge \ldots \mathbf{w}_p = a(\mathbf{u} \wedge \mathbf{w}_2 \wedge \ldots \mathbf{w}_p) + b(\mathbf{w} \wedge \mathbf{w}_2 \wedge \ldots \mathbf{w}_p)$

The vector space generated by these vectors will be denoted by $\bigwedge^p V$. Its dimension is

$$\dim \bigwedge^p V = \binom{n}{p}.$$

We observe that $\bigwedge^p$ and $\bigwedge^{(n-p)}$ have the same dimension and hence isomorphic as vector spaces. Also $\dim \bigwedge^n = 1$.

We can also multiply p-vectors by q-vectors using **Exterior multiplication**

$$\wedge : \bigwedge^p \times \bigwedge^q \to \bigwedge^{(p+q)},$$

$$(\mathbf{w}_1 \wedge \ldots \wedge \mathbf{w}_p) \wedge (\mathbf{u}_1 \wedge \ldots \wedge \mathbf{u}_q) = (\mathbf{w}_1 \wedge \ldots \wedge \mathbf{w}_p \wedge \mathbf{u}_1 \wedge \ldots \wedge \mathbf{u}_q).$$

It is obvious that this exterior multiplication has the following properties:

1. If μ, ν are p and q vectors then $\mu \wedge \nu$ is distributive

2. $\mu \wedge \nu = (-1)^{pq} \nu \wedge \mu$

3. $\mu \wedge (\nu \wedge \chi) = (\mu \wedge \nu) \wedge \chi$

Example 8.1: Let V be the vector space of differentials $dx, dy, dz, \ldots$

1. $(A dx + B dy + C dz) \wedge (D dx + E dy + F dz) = (BF - CE) dy \wedge dz + (CD - AF) dz \wedge dx + (AE - BD) dx \wedge dy$

   It this case the exterior product is "similar" to the vector product operation between two vectors.

2. $(Adx + Bdy + Cdz) \wedge (Ddy \wedge dz + Edz \wedge dx + Fdx \wedge dy) = (AD + BE + CF)dx \wedge dy \wedge dz$

It this case, the exterior product is "similar" to the scalar product operation between two vectors.

8.2 Differential Forms

Let V be the vector space of dimension n.

Definition 8.2: A 1-(differential) form on V is an expression

$$\omega = \sum_{i=1}^{n} a_i dx_i.$$

A p-differential form on V is

$$\omega = \sum_{K} f_K(\mathbf{x})dx_{i_1} \wedge \ldots \wedge dx_{i_p},$$

where K stands for the multi-index $\{i_1, \ldots, i_p\}$ and $f_K(\mathbf{x})$ are smooth functions in $\mathbf{x}$. We observe that a 0-form is just a smooth function of $\mathbf{x}$.

We now want to introduce the concept of external derivative acting on differential forms. We do this first through its definition in R^3.

1. The exterior derivative of a 0-form $\omega = f(\mathbf{x})$ is defined as a 1-form

$$d\omega = \frac{\partial f}{\partial x}dx + \frac{\partial f}{\partial y}dy + \frac{\partial f}{\partial z}dz,$$

2. The exterior derivative of a 1-form

$$\omega = Adx + Bdy + Cdz,$$

is defined as

$$d\omega = \left(\frac{\partial C}{\partial y} - \frac{\partial B}{\partial z}\right)dy \wedge dz + \left(\frac{\partial A}{\partial z} - \frac{\partial C}{\partial x}\right)dz \wedge dx + \left(\frac{\partial B}{\partial x} - \frac{\partial A}{\partial y}\right)dx \wedge dy.$$

3. The exterior derivative of a 2-form

$$\omega = Ady \wedge dz + Bdz \wedge dx + Cdx \wedge dy,$$

is defined as

$$d\omega = \left(\frac{\partial A}{\partial x} + \frac{\partial B}{\partial y} + \frac{\partial C}{\partial z}\right) dx \wedge dy \wedge dz.$$

We see from these definitions that the action of the exterior derivative on a 0-form is equivalent to that of the gradient operator. The action of the exterior derivative on a 1-form($\sim$ vector) is equivalent to that of the curl operator. Finally, the action of the exterior derivative on a two form ($\sim$ axial vector) is equivalent to that of the divergence operator.

We generalize now this operator to p-differential forms.

Let $\mathcal{F}^p$ be the vector space of all p-forms on V.

Definition 8.3: (exterior derivative) d : $\mathcal{F}^p \to \mathcal{F}^{p+1}$ is a mapping which satisfies the following:

1. For a function $f \in \mathcal{F}^0$

$$df = \sum_{i=1}^{n} \frac{\partial f}{\partial x_i} dx_i.$$

2. For $\mu \in \mathcal{F}^p$, $d(d\mu) = 0$. (This is referred to as "Poincare lemma")
3. $d(\omega + \mu) = d\omega + d\mu$
4. $d(\omega \wedge \mu) = d\omega \wedge \mu + (-1)^{deg(\omega)}\omega \wedge d\mu$

From a practical point of view if

$$\omega = \sum_K f_K(\mathbf{x})dx_K, \quad dx_K = dx_{i_1} \wedge \ldots \wedge dx_{i_p},$$

then

$$d\omega = \sum_{K,i} \frac{\partial f_K}{\partial x_i} dx_i \wedge dx_K.$$

Definition 8.4: (1) A differential form $\omega \in \mathcal{F}^p$ is called **exact** if there exists $\mu \in \mathcal{F}^{p-1}$ so that $d\mu = \omega$. (2) A differential form ω is **closed** if $d\omega = 0$.

Example 8.2: Prove that if ω,μ are closed then $\omega \wedge \mu$ is closed

Solution: By the 4th property of the exterior derivative we have

$$d(\omega \wedge \mu) = d\omega \wedge \mu + (-1)^{\deg \omega}\omega \wedge d\mu.$$

Since $d\omega = d\mu = 0$ we have $d(\omega \wedge \mu) = 0$

Example 8.3: Show that if ω is closed and μ is exact then $\omega \wedge \mu$ is exact.

Solution: Since μ is exact there exists ν so that $d\nu = \mu$. Therefore

$$d(\omega \wedge \nu) = d(\omega \wedge \nu = d\omega \wedge d\nu + (-1)^{\deg \omega} \omega \wedge (d\nu) = (-1)^{\deg \omega} \omega \wedge \mu.$$

Hence

$$\omega \wedge \mu = (-1)^{\deg \omega} d(\omega \wedge \nu).$$

i.e $\omega \wedge \mu$ is exact.

Example 8.4: Determine if $\omega = 2yzdx + 2zxdy + 2xydz$ is exact

Solution: It is obvious that $\omega = d(2xyz)$. Therefore ω is closed and exact.

Example 8.5: Determine if $\omega = 2xdx + xydy + yzdz$ is exact

Solution: If ω is exact there exists μ so that $d\mu = \omega$ and therefore $d\omega = d(d\mu) = 0$ However

$$d\omega = ydx \wedge dy + zdy \wedge dz \neq 0,$$

therefore ω is not closed nor exact.

Example 8.6: Is it true that $\omega \wedge d\omega = 0$

Solution: The answer is NO. For example, if $\omega = ydx + dz$ then $d\omega = dy \wedge dx$ and therefore

$$\omega \wedge d\omega = dy \wedge dx \wedge dz.$$

We now list the properties of the exterior derivatives under mappings

Theorem 8.1: Let $\psi : M \to N$ and let ω, μ be two differential forms on N then

1. $\psi^*(\omega + \mu) = \psi^*(\omega) + \psi^*(\mu)$
2. $\psi^*(\omega \wedge \mu) = \psi^*(\omega) \wedge \psi^*(\mu)$
3. $d(\psi^*(\omega)) = \psi^*(d\omega)$

To prove the third property we consider a one form on N which can be written locally as

$$df = \sum \frac{\partial f}{\partial y_j} dy_j.$$

Since $dy_j = \sum \frac{\partial y_j}{\partial x_i} dx_i$ we have

$$\psi^*(dg) = \sum \frac{\partial f(\mathbf{y(x)})}{\partial y_j} \frac{\partial y_j}{\partial x_i} dx_i = \sum \frac{\partial (\psi^* f)}{\partial x_i} dx_i = d(\psi^* f),$$

and this can be generalized to higher order forms.

8.3 Hodge Star Operator

Definition 8.5: Let V be a vector space. An inner product on V is a mapping

$$(*,*) : V \times V \to R,$$

so that

1. $(\mathbf{v}, \mathbf{w}) = 0$ for all $\mathbf{w}$ implies that $\mathbf{v} = \mathbf{0}$
2. $(\mathbf{v}, \mathbf{w}) = (\mathbf{w}, \mathbf{v})$
3. $(*,*)$ is linear in both variables

Observe that we do not require that $(\mathbf{v}, \mathbf{w}) \geq 0$ to accommodate vector spaces with nonpositive metrics.

Example 8.7: The inner product on Minkoviski space is

$$(v, w) = v_1 w_1 + v_2 w_2 + v_3 w_3 - v_4 w_4,$$

where $\mathbf{v} = (v_1, v_2, v_3, v_4)$ and $\mathbf{w} = (w_1, w_2, w_3, w_4)$ (and the velocity of light has been normalized to 1.)

An orthonormal basis of V is a set of vectors $\{\mathbf{v_1}, \ldots, \mathbf{v_n}\}$ so that

$$(\mathbf{v_i}, \mathbf{v_j}) = \pm \delta_{ij}.$$

Once again observe that we allow some of the basis vectors to have "negative length".

Theorem 8.2: Let f be a linear functional on V. There exist a unique vector $\mathbf{w} \in V$ so that

$$f(\mathbf{v}) = (\mathbf{v}, \mathbf{w}), \quad \text{for all } \mathbf{v} \in V.$$

For a proof see the Appendix.

Assuming that V is endowed with a scalar product we now define "an induced" scalar product on $\bigwedge^p V$.

Definition 8.6: Let $\omega = \mathbf{v_1} \wedge \ldots \wedge \mathbf{v}_p$ and $\mu = \mathbf{w_1} \wedge \ldots \wedge \mathbf{w}_p$ we define the scalar product

$$(\omega, \mu) = \det(G),$$

where G is the Gramm matrix whose entries are $G_{ij} = (\mathbf{v}_i, \mathbf{w}_j)$.

Example 8.8: Let $V = \{dx^1, dx^2, dx^3\}$ and define a scalar product on V by $(dx^i, dx^j) = \delta_{ij}$. To see how the scalar product on $V \wedge V$ "looks like" we observe that a basis of this vector space is $\{dx^1 \wedge dx^2, dx^2 \wedge dx^3, dx^3 \wedge dx^1\}$. Using the definition above we have

$$(\ dx^1 \wedge dx^2, \quad dx^2 \wedge dx^3 \ ) = \begin{vmatrix} (dx^1, dx^2) & (dx^1, dx^3) \\ (dx^1, dx^2) & (dx^2, dx^3) \end{vmatrix} = \begin{vmatrix} 0 & 0 \\ 1 & 0 \end{vmatrix} = 0. \ (8.1)$$

Similarly $\{dx^2 \wedge dx^3, dx^3 \wedge dx^1) = (dx^3 \wedge dx^1, dx^1 \wedge dx^2) = 0$ but $(dx^1 \wedge dx^2, dx^1 \wedge dx^2) = 1$ etc.

Remark 8.1: The vector space $\bigwedge^p V$ has dimension $\frac{n!}{p!(n-p)!}$. Therefore $\bigwedge^p V$ and $\bigwedge^{n-p} V$ have the same dimension and therefore isomorphic. Furthermore $\bigwedge^n V$ has dimension 1 and is isomorphic to R.

For any $\omega \in \bigwedge^p$ we define now a functional

$$\omega : \overset{n-p}{\bigwedge} V \to \overset{n}{\bigwedge} V,$$

by

$$\omega(\mu) = \omega \wedge \mu \in \overset{n}{\bigwedge} V.$$

By theorem proved in the appendix there exists an element $*\omega \in \bigwedge^{n-p} V$ so that

$$\omega(\mu) = \omega \wedge \mu = (*\omega, \mu).$$

The mapping $\omega \to *\omega$ is called the **Hodge Star Mapping**. It represents the isomorphism between $\bigwedge^n V$ and $\bigwedge^{n-p} V$.

Example 8.9: Let $V = \{dx^1, dx^2, dx^3\}$. Then

$$V \wedge V = \{dx^1 \wedge dx^2, dx^2 \wedge dx^3, dx^3 \wedge dx^1\}.$$

Therefore V is isomorphic to $V \wedge V$ (Both have dimension 3).

How the functional dx^1 acts on $V \wedge V$? It is enough to determine its action on the basis elements of $V \wedge V$,

By definition we have

$$dx^1(dx^1 \wedge dx^2) = dx^1 \wedge dx^1 \wedge dx^2 = 0, \quad dx^1(dx^2 \wedge dx^3) = dx^1 \wedge dx^2 \wedge dx^3 = 1,$$

$$dx^1(dx^3 \wedge dx^1) = dx^1 \wedge dx^3 \wedge dx^1 = 0.$$

What is $*dx^1$? We are seeking an element of $V \wedge V$ whose scalar product with the elements of this space is exactly the same as the action of the functional dx^1. Obviously

$$*dx^1 = dx^2 \wedge dx^3.$$

In fact

$$(dx^2 \wedge dx^3, dx^1 \wedge dx^2) = 0, \quad (dx^2 \wedge dx^3, dx^2 \wedge dx^3) = 1, \quad (dx^2 \wedge dx^3, dx^3 \wedge dx^1) = 0.$$

We now seek an explicit representation of the isomorphism between $\bigwedge^p V$ and $\bigwedge^{(n-p)} V$ when V is endowed with aa scalar product.

Consider a fixed element Let $\omega \in \bigwedge^p V$ and let $\mu \in \bigwedge^{(n-p)} V$. We define a linear mapping

$$\overset{(n-p)}{\bigwedge} V \to \overset{n}{\bigwedge} V, \quad \text{by } \mu \to \omega \wedge \mu.$$

Since $\bigwedge^n V$ is one-dimensional and therefore isomorphic to R this mapping defines a functional on $\bigwedge^{(n-p)} V$. By the theorem quoted above about such a functional there exist $*\omega \in \bigwedge^{(n-p)} V$ so that

$$\omega \wedge \mu = (*\omega, \mu)\sigma,$$

where $\sigma = \sigma_1 \wedge \ldots \wedge \sigma_n$ and $\{\sigma_1, \ldots, \sigma_n\}$ is an orthonormal basis of $\bigwedge^n V$. This operator $* : \omega \to *\omega$ is called the Hodge star operator (or mapping).

To obtain a concrete representation of this mapping for a given orthonormal basis $\{\sigma_1, \ldots, \sigma_n\}$ of $\bigwedge^n V$ we consider the case $\omega = \sigma_1 \wedge \ldots \wedge \sigma_p$. If H is a multi-index of length $(n-p)$ then

$$\omega \wedge \sigma^H = (*\omega, \sigma^H)\sigma.$$

The left-hand side of this equation is zero unless $H = \{p+1, \ldots, n\}$ hence

$$*\omega = c\sigma^{p+1} \wedge \ldots \wedge \sigma^n.$$

To determine c we let $H = (p+1, \ldots, n)$. We then have

$$\omega \wedge \sigma^H = \sigma = c(\sigma^H, \sigma^H)\sigma,$$

therefore $c = (\sigma^H, \sigma^H) = \pm 1$ and

$$*\omega = (\sigma^H, \sigma^H)\sigma^H.$$

Theorem 8.3:

1. If ω is a p-vector then

$$**\omega = (-1)^{p(n-p)+(n-s)/2}\omega.$$

2. If ω, μ are p-vectors then

$$\omega \wedge *\mu = \mu \wedge *\omega = (-1)^{(n-s)/2}(\omega, \mu)\sigma.$$

In these formulas s is the signature of the orthonormal basis of V i.e $s = p - m$ where p the number of vectors σ_i of length 1 and m is the number of vectors in the basis whose length is -1.

Example 8.10: In R^n with the regular Euclidean metric $c = 1$ and $s = 3$ therefore if f, g are two functions then

1. $df = \frac{\partial f}{\partial x}dx + \frac{\partial f}{\partial y}dy + \frac{\partial f}{\partial z}dz$

2. $*df = \frac{\partial f}{\partial x}dy \wedge dz + \frac{\partial f}{\partial y}dz \wedge dx + \frac{\partial f}{\partial z}dx \wedge dy$

3. $df \wedge *dg = \left(\frac{\partial f}{\partial x}\frac{\partial g}{\partial x} + \frac{\partial f}{\partial y}\frac{\partial g}{\partial y} + \frac{\partial f}{\partial z}\frac{\partial g}{\partial z}\right) dx \wedge dy \wedge dz$

8.3.1 Inverse of Poincare Lemma

In vector calculus, one proves that a vector field $\mathbf{v}$ whose curl is zero is conservative and there exist a potential ϕ so that $grad\phi = \mathbf{v}$. In the language of differential forms this is equivalent to saying that if the exterior derivative of a one form

$$\mathbf{v} = v_1 dx + v_2 dy + v_3 dz,$$

is zero then there exists a zero form ϕ so that $d\phi = \mathbf{v}$.

Similarly, one proves in vector calculus that if the divergence of a vector field $\mathbf{w}$ is zero then there exists a vector potential $\mathbf{A}$ so that $curl\mathbf{A} = \mathbf{w}$. In the language of differential forms this can be reformulated to say that if the exterior derivative of a two-form

$$\mathbf{w} = w_1 dy \wedge dz + w_2 dz \wedge dx + w_3 dx \wedge dy,$$

is zero then there exist a one form $\mathbf{A}$ so that $d\mathbf{A} = \mathbf{w}$

The inverse of Poincare lemma (which says that for any differential form ω, $d(d\omega) = 0$ states that the previous two examples can be generalized.

Theorem 8.4: Let $\omega \in \mathcal{F}^p$ be closed i.e $d\omega = 0$. There exists $\mu \in \mathcal{F}^{p-1}$ so that $d\mu = \omega$.

We observe that μ is not unique in fact if $\nu \in \mathcal{F}^{p-2}$ then

$$d(\mu + d\nu) = d\mu + d(d\nu) = d\mu = \omega.$$

8.4 Applications

A. Derive a necessary condition for the integrability of the system

$$\frac{\partial R}{\partial y} - \frac{\partial Q}{\partial z} = f(\mathbf{x}), \quad \frac{\partial P}{\partial z} - \frac{\partial R}{\partial x} = g(\mathbf{x}), \quad \frac{\partial Q}{\partial x} - \frac{\partial P}{\partial y} = h(\mathbf{x}).$$

Solution: In the language of differential forms, we can reformulate this problem as follows:

Let be given the second-order form

$$\omega = f(\mathbf{x})dydz + g(\mathbf{x})dzdx + h(\mathbf{x})dxdy,$$

find a first order form

$$\mu = Pdx + Qdy + Rdz,$$

so that $d\mu = \omega$. However, this relation implies that $d\omega = 0$ therefore a necessary condition for the integrability of the system above is that

$$\frac{\partial f}{\partial x} + \frac{\partial g}{\partial y} + \frac{\partial h}{\partial z} = 0.$$

Observe that this condition is also sufficient by the inverse of Poincare lemma.

B. Let $H(\mathbf{x}, \mathbf{y})$ be a homogeneous function of degree two in $\mathbf{y}$. Under a coordinate transformation

$$(\mathbf{x}, \mathbf{y}) \to (\mathbf{x}, \mathbf{p}), \quad \mathbf{p} = \left(\frac{\partial H}{\partial y_1}, \dots, \frac{\partial H}{\partial y_n} \right).$$

H becomes

$$H(\mathbf{x}, \mathbf{y}) = S(\mathbf{x}, \mathbf{p}).$$

Show that

$$\frac{\partial S}{\partial x_i} = -\frac{\partial H}{\partial x_i}, \quad \frac{\partial S}{\partial p_i} = y_i.$$

Since H is homogeneous of degree two in $\mathbf{y}$ we have

$$2H = \sum \frac{\partial H}{\partial x_i} y_i = \sum p_i y_i.$$

Therefore

$$2dH = \sum p_i dy_i + \sum y_i dp_i. \tag{8.2}$$

On the other hand from the definition of H we have

$$dH = \sum \frac{\partial H}{\partial x_i} dx_i + \sum \frac{\partial H}{\partial y_i} dy_i = \sum \frac{\partial H}{\partial x_i} dx_i + \sum p_i dy_i. \tag{8.3}$$

Subtracting (8.3) from (8.2) we have

$$dH = -\sum \frac{\partial H}{\partial x_i} dx_i + \sum y_i dp_i.$$

However for S we have

$$dS = \sum \frac{\partial S}{\partial x_i} dx_i + \sum \frac{\partial S}{\partial p_i} dp_i.$$

Since $H = S$ we have $dH = dS$ and we have the desired results.

8.5 Differential Geometry of Manifolds in R^3

In in this section, we apply the theory of differential forms to derive the basic results about surfaces in R^3.

8.5.1 Frames in R^3

Introducing a fixed Cartesian coordinate system (x, y, z) in R^3 we attach to each point in R^3 an orthogonal (right handed) coordinate system with unit vectors e_1, e_2, e_3. (This will be referred to as a frame). We assume that the orientation of these vectors as function of the position is smooth.

The differentials $\mathbf{dx} = (dx, dy, dz)$ can be expressed in terms of local frame as

$$\mathbf{dx} = dx\mathbf{i} + dy\mathbf{j} + dz\mathbf{k} = \sum \mu_i e_i. \tag{8.4}$$

Furthermore we have

$$de_i = \sum \gamma_{ij} e_j. \tag{8.5}$$

In these expressions μ_i, γ_{ij} are one forms.

From the orthogonality of the frame vectors $e_i \cdot e_j = \delta_{ij}$ we obtain

$$d(e_i) \cdot e_j + e_i \cdot d(e_j) = 0.$$

Hence using (8.5) we have

$$\left(\sum_k \gamma_{ik} e_k \right) \cdot e_j + e_i \cdot \left(\sum_k \gamma_{jk} e_k \right),$$

which yields

$$\gamma_{ij} + \gamma_{ji} = 0.$$

(and $\gamma_{ii} = 0$). Hence we can write

$$\Gamma = \begin{pmatrix} 0 & \gamma_{12} & \gamma_{13} \\ -\gamma_{12} & 0 & \gamma_{23} \\ -\gamma_{13} & -\gamma_{23} & 0 \end{pmatrix}. \tag{8.6}$$

(Observe that $\Gamma = -\Gamma^T$). In vector notation these results can be written as

$$\mathbf{dx} = \mu e = \mu_1 e_1 + \mu_2 e_2 + \mu_3 e_3, \tag{8.7}$$

$$de = \Gamma e, \tag{8.8}$$

where $\boldsymbol{\mu} = (\mu_1, \mu_2, \mu_3)$ and $\mathbf{e}^T = (\mathbf{e}_1, \mathbf{e}_2, \mathbf{e}_3)$. Applying the exterior derivatives to (8.7) we obtain

$$0 = d(d\mathbf{x}) = d\boldsymbol{\mu}\,\mathbf{e} - \boldsymbol{\mu} \wedge d\mathbf{e}.$$

Taking the wedge product of (8.8) by $\boldsymbol{\mu}$ and using the previous result leads to

$$\boldsymbol{\mu} \wedge d\mathbf{e} = \boldsymbol{\mu} \wedge \Gamma \mathbf{e} \qquad (8.9)$$
$$= d\boldsymbol{\mu}\,\mathbf{e}.$$

Therefore

$$d\boldsymbol{\mu} - \boldsymbol{\mu} \wedge \Gamma = 0. \qquad (8.10)$$

Similarly by applying the exterior derivative to (3.24) yields

$$0 = d(d\mathbf{e}) = d\Gamma\mathbf{e} - \Gamma \wedge d\mathbf{e}. \qquad (8.11)$$

Using (8.5) we obtain

$$d\Gamma = \Gamma \wedge \Gamma. \qquad (8.12)$$

Equations (8.6)–(8.8) are referred to as the structure equations of the frame while Eqs. (8.10) and (8.12) are the integrability conditions.

Example 8.11: Spherical Coordinates

In Spherical coordinates

$$\mathbf{x} = (r\sin\phi\cos\theta, r\sin\phi\sin\theta, r\cos\phi).$$

Hence

$$dx = (\sin\phi\cos\theta, \sin\phi\sin\theta, \cos\phi)dr \qquad (8.13)$$
$$+ (r\cos\phi\cos\theta, r\cos\phi\sin\theta, -r\sin\phi)d\phi$$
$$+ (-r\sin\phi\sin\theta, r\sin\phi\cos\theta, 0)d\theta.$$

We can write therefore

$$d\mathbf{x} = (dr)\mathbf{e}_1 + (rd\phi)\mathbf{e}_2 + (r\sin\phi d\theta)\mathbf{e}_3,$$

ie.

$$\mu_1 = dr, \quad \mu_2 = rd\phi, \quad \mu_3 = r\sin\phi d\theta.$$

Since

$$\mathbf{e}_1 = (\sin\phi\cos\theta, \sin\phi\sin\theta, \cos\phi).$$

we have

$$de_1 = (\cos\phi\cos\theta, \cos\phi\sin\theta, -\sin\phi)d\phi + (-\sin\phi\sin\theta, \sin\phi\cos\theta, 0)d\theta$$
$$= (d\phi)\mathbf{e}_2 + (\sin\phi d\theta)\mathbf{e}_3.$$

Similarly we obtain

$$d\mathbf{e}_2 = (-d\phi)\mathbf{e}_1 + (\cos\phi d\theta)\mathbf{e}_3.$$

Hence the matrix Γ is

$$\Gamma = \begin{pmatrix} 0 & d\phi & \sin\phi d\theta \\ -d\phi & 0 & \cos\phi d\theta \\ -\sin\phi d\theta & -\cos\phi d\theta & 0 \end{pmatrix}. \qquad (8.14)$$

8.5.2 Smooth Manifolds in R^3

Let S be a smooth surface in R^3. At each point $\mathbf{x}$ on S we choose a frame so that $\mathbf{e}_1$ and $\mathbf{e}_2$ are in the tangent plane to S at $\mathbf{x}$ while $\mathbf{e}_3$ is in the normal direction. If we constrain the motion of this moving frame to S, we have

$$d\mathbf{x} = \mu_1 \mathbf{e}_1 + \mu_2 \mathbf{e}_2, \qquad (8.15)$$

(i.e. $\mu_3 = 0$). The integrability equations (8.10) and (8.12) of a frame constrained to S become

$$d\mu_1 = \gamma_{12} \wedge \mu_2, \ \ d\mu_2 = -\gamma_{12} \wedge \mu_1, \ \ \mu_1 \wedge \gamma_{13} + \mu_2 \wedge \gamma_{23} = 0, \qquad (8.16)$$

$$d\gamma_{12} = -\gamma_{13} \wedge \gamma_{23}, \ \ d\gamma_{13} = \gamma_{12} \wedge \gamma_{23}, \ \ d\gamma_{23} = -\gamma_{12} \wedge \gamma_{13}. \qquad (8.17)$$

The first equation in (8.17) implies that γ_{12} is determined by γ_{13} and γ_{23}. Furthermore since there is only one independent 2-form on two dimensional space we infer that

$$\gamma_{13} \wedge \gamma_{23} = K\mu_1 \wedge \mu_2, \qquad (8.18)$$

K is a scalar function which is called the **Gaussian curvature** of S. Another two form is $\mu_1 \wedge \gamma_{23} - \mu_2 \wedge \gamma_{13}$. Hence, in this case also, we must have

$$\mu_1 \wedge \gamma_{23} - \mu_2 \wedge \gamma_{13} = -2H\mu_1 \wedge \mu_2. \qquad (8.19)$$

H is referred to as the **mean curvature** of S.

Example 8.12: The equation for the two dimensional torus T^2 (surface in three dimensions is

$$x(u^1, u^2) = ((a + b\cos u^2)\cos u^1, (a + b\cos u^2)\sin u^1, b\sin u^2).$$

Hence

$$d\mathbf{x} = [-(a + b\cos u^2)\sin u^1), (a + b\cos u^2)\cos u^1, 0]du^1$$
$$+ [-b\sin u^2 \cos u^1, -b\sin u^2 \sin u^1, b\cos u^2]du^2.$$

Normalizing the vectors in this formula we have

$$d\mathbf{x} = [-\sin u^1, \cos u^1, 0](a + b\cos u^2)du^1$$
$$+ [-\sin u^2 \cos u^1, -\sin u^2 \sin u^1, \cos u^2](rdu^2).$$

Since the vectors in this formula are orthonormal, we have

$$\mathbf{e}_1 = [-\sin u^1, \cos u^1, 0], \ \ \mathbf{e}_2 = [-\sin u^2 \cos u^1, -\sin u^2 \sin u^1, \cos u^2],$$

and

$$\mu_1 = (a + b\cos u^2)du^1, \ \ \mu_2 = bdu^2.$$

Taking the vector product of e_1 and e_2 we obtain

$$e_3 = e_1 \times e_2 = [\cos u^1 \cos u^2, \sin u^1 \cos u^2, \sin u^2].$$

we next compute de_3 and express it in terms of e_1, e_2 and use (3.24) to evaluate γ_{13} and γ_{23}

$$\begin{aligned} de_3 &= [-\sin u^1 \cos u^2, \cos u^1 \cos u^2, 0]du^1 \\ &\quad - [\sin u^2 \cos u^1, -\sin u^2 \sin u^1, cosu^2)]du^2 \\ &= -\gamma_{13}e_1 + -\gamma_{23}e_2. \end{aligned}$$

Solving this system of equation we obtain

$$\gamma_{13} = -\cos u^2 du^1, \quad \gamma_{23} = -du^2.$$

Similarly we have

$$de_1 = \gamma_{12}e_2 + \gamma_{13}e_3.$$

Solving this equation for γ_{12} yields

$$\gamma_{12} = \sin u^2 du^1.$$

Using (8.18) and (8.19), we obtain the following for the Gaussian and mean curvature of the torus

$$K = \frac{\cos u^2}{b(a + b\cos u^2)}, \quad 2H = -\frac{a + 2b\cos u^2}{b(a + b\cos u^2)}.$$

There is also a (rather complicated) closed formula for Gauss curvature of a surface x in terms of the metric tensor g_{ij}.

Since a two-dimensional tangent plane has only two independent one forms, we can write

$$\gamma_{13} = a\mu_1 + b\mu_2, \quad \gamma_{23} = c\mu_1 + d\mu_2. \tag{8.20}$$

However using the last equation in (8.16) yields $b = c$. Inserting these relations in (8.18),(8.19), we obtain

$$K = ad - b^2, \quad H = \frac{a + d}{2}. \tag{8.21}$$

The eigenvalues λ_1, λ_2 of the matrix

$$A = \begin{pmatrix} a & b \\ b & d \end{pmatrix}, \tag{8.22}$$

are called the **principal curvatures** of S, and it is easy to see that

$$K = \lambda_1\lambda_2, \quad H = \frac{\lambda_1 + \lambda_2}{2}.$$

Using the first equation in (8.17) and (8.18) leads to

$$d\gamma_{12} + K\mu_1\mu_2 = 0. \tag{8.23}$$

However $\gamma_{12} = \alpha\mu_1 + \beta\mu_2$ and the first two equations of (8.17) can the be used to determine γ_{12} if μ_1 and μ_2 are known. Hence (8.23) implies that K is determined completely in terms of μ_1 and μ_2. Thus K is an intrinsic invariant of S.

Theorem 8.5 (Gauss) K and H (viz.the Gaussian and Mean curvature of a manifold) are geometric invariants which are independent of the parametric representation of the manifold.

Example 8.13: Compute the Gaussian and Mean curvatures of the ellipsoid.

$$\frac{x^2}{a^2} + \frac{y^2}{a^2} + \frac{z^2}{c^2} = 1.$$

Solution: A parameterization of the given ellipsoid is

$$\mathbf{x}(\theta, \phi) = (a\sin\phi\cos\theta, a\sin\phi\sin\theta, c\cos\phi).$$

For a motion on the surface, we therefore have

$$d\mathbf{x} = \frac{(a\cos\phi\cos\theta, a\cos\phi\sin\theta, -c\sin\phi)}{\sqrt{a^2\cos^2\phi + c^2\sin^2\phi}}\left(\sqrt{a^2\cos^2\phi + c^2\sin^2\phi}\,d\phi\right)$$
$$+(-\sin\theta, \cos\theta, 0)(a\sin\phi\,d\theta).$$

Hence the two unit length tangent vectors on the surface are

$$\mathbf{e}_1 = \frac{(a\cos\phi\cos\theta, a\cos\phi\sin\theta, -c\sin\phi)}{\sqrt{a^2\cos^2\phi + c^2\sin^2\phi}}, \quad \mathbf{e}_2 = (-\sin\theta, \cos\theta, 0).$$

It follows then that

$$\mu_1 = \sqrt{a^2\cos^2\phi + c^2\sin^2\phi}\,d\phi, \quad \mu_2 = a\sin\phi\,d\theta.$$

The vector $\mathbf{e}_3$ which is normal to $\mathbf{e}_1, \mathbf{e}_2$ is

$$\mathbf{e}_3 = \frac{1}{\sqrt{a^2\cos^2\phi + c^2\sin^2\phi}}(c\sin(\phi)\cos(\theta), c\sin(\phi)\sin(\theta), a\cos(\phi)).$$

Computations similar to those done in the previous example about the torus yield

$$\gamma_{13} = -\frac{ca\,d\phi}{a^2\cos(\phi)^2 + c^2\sin(\phi)^2}, \quad \gamma_{23} = -\frac{c\sin(\phi)\,d\theta}{\sqrt{a^2\cos(\phi)^2 + c^2\sin(\phi)^2}},$$

and

$$\gamma_{12} = \frac{a\cos(\phi)d\theta}{\sqrt{a^2\cos(\phi)^2 + c^2\sin(\phi)^2}}.$$

Using (8.18) and (8.19), we obtain

$$K = \frac{c^2}{(a^2\cos(\phi)^2 + c^2\sin(\phi)^2)^2},$$

$$H = -\frac{1}{2}\frac{c(a^2\cos(\phi)^2 + c^2\sin(\phi)^2 + a^2)}{a(a^2\cos(\phi)^2 + c^2\sin(\phi)^2)^{3/2}}.$$

We observe that when $c = a = R$ (case of a sphere) $K = \frac{1}{R^2}$ and $H = \frac{1}{R}$.

8.5.3 The Laplace Operator

In R^3

$$df = \frac{\partial f}{\partial x}dx + \frac{\partial f}{\partial y}dy + \frac{\partial f}{\partial z}dz.$$

Therefore

$$*df = \frac{\partial f}{\partial x}dy \wedge dz + \frac{\partial f}{\partial y}dz \wedge dx + \frac{\partial f}{\partial x}dx \wedge dy.$$

Hence

$$d(*df) = \left(\frac{\partial^2 f}{\partial x^2} + \frac{\partial^2 f}{\partial y^2} + \frac{\partial^2 f}{\partial z^2}\right)dx \wedge dy \wedge dz.$$

That is the Laplace operator can be identified with the operator $d(*d)$ acting on a function f.

Using this observation, it is possible to generalize this definition to any smooth manifold in R^3. On such a manifold, there are only two independent first-order differential forms μ_1, μ_2 and

$$d\mathbf{x} = \mu_1 \mathbf{e}_1 + \mu_2 \mathbf{e}_2.$$

Since $\mathbf{e}_1$, $\mathbf{e}_2$ are orthogonal and $\mathbf{e}_i \cdot \mathbf{e}_j = \delta_{ij}$ we define on the vector space $V = \{\mu_1, \mu_2\}$ a scalar product by

$$\boldsymbol{\mu}_i \cdot \boldsymbol{\mu}_j = \delta_{ij}.$$

If $\omega \in V$ then $\omega = a\mu_1 + b\mu_2$ and the action of ω (when considered as an element of $V \wedge V \cong V$) on the basis is

$$\omega(\mu_1) = (a\mu_1 + b\mu_2) \wedge \mu_1 = -b\mu_1 \wedge \mu_2, \quad \omega(\mu_2) = (a\mu_1 + b\mu_2) \wedge \mu_2 = a\mu_1 \wedge \mu_2.$$

Hence

$$(*\omega, \mu_1) = -b, \quad (*\omega, \mu_2) = a,$$

that is

$$*\omega = -b\mu_1 + a\mu_2.$$

In the particular case where f(u,v) is a function on a manifold R^3 and

$$\omega = df = \frac{\partial f}{\partial u}du + \frac{\partial f}{\partial v}dv,$$

then $a = \frac{\partial f}{\partial u}$, $b = \frac{\partial f}{\partial v}$ and

$$*\omega = *df = -\frac{\partial f}{\partial v}du + \frac{\partial f}{\partial u}dv.$$

Hence

$$d(*df) = -\frac{\partial^2 f}{\partial v^2}dv \wedge du + \frac{\partial^2 f}{\partial u^2}du \wedge dv = (\frac{\partial^2 f}{\partial u^2} + \frac{\partial^2 f}{\partial v^2})du \wedge dv = (\Delta f)du \wedge dv.$$
$$(8.24)$$

Remark 8.2: In terms of the metric tensor g_{ij}, the Laplace operator on a manifold in R^n is defined by

$$\nabla^2 f = \frac{1}{\sqrt{|g|}}\partial_i \left(\sqrt{|g|}g^{ij}\partial_j f\right).$$

This formula applies to Riemann and Pseudo-Riemann manifolds.

For the position vector $\mathbf{x}$ on a manifold we have from (8.15)

$$d\mathbf{x} = \mu_1\mathbf{e}_1 + \mu_2\mathbf{e}_2, \quad *d\mathbf{x} = \mu_2\mathbf{e}_1 - \mu_1\mathbf{e}_2 = d\mathbf{x} \times \mathbf{e}_3,$$

where $\mathbf{e}_3 = \mathbf{e}_1 \times \mathbf{e}_2$. Using (8.5), (8.15) and (8.19) yields.

$$d(*d\mathbf{x}) = d[d\mathbf{x} \times \mathbf{e}_3] = -d\mathbf{x} \times d\mathbf{e}_3 = -d\mathbf{x} \times d[\mathbf{e}_1 \times \mathbf{e}_2] = -2H(\mu_1 \wedge \mu_2)\mathbf{e}_3,$$

i.e.

$$\Delta\mathbf{x} = (\Delta x, \Delta y, \Delta z) = -2H\mathbf{e}_3,$$

where H is the mean curvature of the manifold. This motivates the following

Definition 8.7: A surface S is called minimal if its mean curvature is zero.

Corollary 8.1: For a minimal surface, the coordinate functions are harmonic.

Example 8.14: The catenoid (which is obtained by revolving a a catenary around its axis is a minimal surface. The parametric equation of this surface is

$$\mathbf{x} = (\cosh(u)\cos v, \cosh(u)\sin v, u).$$

Solution: For this case

$$\Delta f = \frac{\partial^2 f}{\partial u^2} + \frac{\partial^2 f}{\partial v^2}.$$

and it is straightforward to verify that $\Delta x = \Delta y = \Delta z = 0$.

8.5.4   Maxwell Equations in Free Space

In Free space Maxwell Equation reduce to to two vector equations, one for the Electric field $\mathbf{E} = (E_1, E_2, E_3)$ and the second for the magnetic field $\mathbf{H} = (H_1, H_2, H_3)$.

$$\nabla \times \mathbf{E} = -\frac{1}{c}\frac{\partial \mathbf{H}}{\partial t},$$

$$\nabla \times \mathbf{H} = \frac{1}{c}\frac{\partial \mathbf{E}}{\partial t}.$$

where c is the speed of light. Since these equations are invariant under Lorenz transformations, we use Minkowiski space-time coordinates that satisfy

$$ds^2 = dx_1^2 + dx_2^2 + dx_3^2 - c^2 dt^2,$$

and the differentials as one-forms satisfy the orthogonality relations

$$(dx_i, dx_j) = \delta_{i,j}, \quad (dx_i, cdt) = 0, \quad (cdt, cdt) = -1.$$

This implies that

$$*(dx_1 \wedge dx_2) = -dx_3 \wedge (cdt), \quad *(dx_2 \wedge dx_3) = -dx_1 \wedge (cdt), \quad *(dx_3 \wedge dx_1) = -dx_2 \wedge (cdt),$$

$$*(dx_1 \wedge (cdt)) = dx_2 \wedge dx_3, \quad *(dx_2 \wedge (cdt)) = dx_3 \wedge dx_2, \quad *(dx_3 \wedge (cdt)) = dx_1 \wedge dx_2.$$

To express Maxwell equations in terms of differential forms we introduce

$$\gamma = (E_1 dx_1 + E_2 dx_2 + E_3 dx_3) \wedge (cdt) + (H_1 dx_2 \wedge dx_3 + H_2 dx_3 \wedge dx_1 + H_3 dx_1 \wedge dx_2).$$

Hence

$$*\gamma = (E_1 dx_2 \wedge dx_3 + E_2 dx_3 \wedge dx_1 + E_3 dx_1 \wedge dx_2) - (H_1 dx_1 + H_2 dx_2 + H_3 dx_3) \wedge (cdt).$$

Maxwell equations for $\mathbf{E}$ and $\mathbf{H}$ in free space can be expressed now as

$$d\gamma = 0, \quad d*\gamma = 0.$$

Appendix 8A: Functionals over a Vector Space

Let V be a vector space over R.

   Definition 8.8: A linear functional on V is a mapping

$$f : V \to R,$$

with the following properties,

1. $f(\mathbf{v_1} + \mathbf{v_2}) = f(\mathbf{v_1}) + f(\mathbf{v_2})$
2. $f(a\mathbf{v}) = af(\mathbf{v}), \quad a \in R$

Example 8.15: If $\mathbf{v} = (\alpha_1, \ldots, \alpha_n)$ then

$$f(\mathbf{v}) = a_1\alpha_1 + \cdots + a_n\alpha_n, \quad a_i \in R,$$

is a linear functional on V.

We consider now the set of all functionals on V and define

$$F = \{f \,|\, f \text{ is a linear functional on } V\}.$$

Multiplication of a functional by a scalar and addition of functionals are defined as follows:

1. $(af)(\mathbf{v}) = af(\mathbf{v})$
2. $(f_1 + f_2)(\mathbf{v}) = f_1(\mathbf{v}) + f_2(\mathbf{v})$.

With these definitions F is a vector space over R.

Theorem 8.6: Let $f \in F$. The action of f on any vector $\mathbf{v} \in V$ is determined by its action on a basis $\{\mathbf{v_1}, \ldots, \mathbf{v_n}\}$ of V.

Proof: Suppose that $f(\mathbf{v}_i) = \alpha_i$ are known then for any vector $\mathbf{v}$ we have

$$\mathbf{v} = a_1\mathbf{v_1} + \cdots + a_n\mathbf{v_n}.$$

Therefore

$$f(\mathbf{v}) = f(a_1\mathbf{v_1} + \cdots + a_n\mathbf{v_n}) = a_1 f(\mathbf{v_1}) + \cdots + a_n f(\mathbf{v_n}) = a_1\alpha_1 + \cdots + a_n\alpha_n.$$

∎

We now define the following set of functionals in F

Definition 8.9: $f_i(\mathbf{v}_j) = \delta_{ij}, \quad i, j = 1, \ldots, n$

By the previous theorem these functionals are well defined since their values on the basis $\{\mathbf{v_1}, \ldots, \mathbf{v_n}\}$ has been specified.

Theorem 8.7: $\{f_1, \ldots, f_n\}$ is a basis of F.

Proof: It is obvious that $\{f_1, \ldots, f_n\}$ is an independent set. We now show that any $f \in F$ can be expressed as a linear combination of these functionals.

Let f be a functional in F and suppose that $f(\mathbf{v}_i) = a_i$. We now claim that $f = a_1 f_1 + \ldots + a_n f_n$. In fact if $\mathbf{v} = (\alpha_1, \ldots, \alpha_n)$ then

$$f(\mathbf{v}) = \alpha_1 f(\mathbf{v_1}) + \ldots + \alpha_n f(\mathbf{v_n}) = \alpha_1 a_1 + \cdots + \alpha_n a_n.$$

On the other hand

$$
\begin{aligned}
(a_1 f_1 + \cdots + a_n f_n)(\mathbf{v}) &= a_1 f_1(\mathbf{v}) + \cdots + a_n f_n(\mathbf{v}) \\
&= a_1 f_1(\alpha_1 \mathbf{v}_1 + \cdots + \alpha_n \mathbf{v}_n) + \cdots \\
&\quad + a_n f_n(\alpha_1 \mathbf{v}_1 + \cdots + \alpha_n \mathbf{v}_n) \\
&= \alpha_1 a_1 + \cdots + \alpha_n a_n.
\end{aligned}
\tag{8A.1}
$$

which demonstrates that $f = a_1 f_1 + \cdots + a_n f_n$.

Definition 8.10: $\{f_1, \ldots, f_n\}$ is called the the dual basis in F to $\{\mathbf{v}_1, \ldots, \mathbf{v}_n\}$ in V.

Definition 8.11: A scalar product on V is a mapping

$$
(*, *) : V \times V \to R \quad \text{so that.}
$$

1. The mapping is linear in each argument.
2. $(\mathbf{v}, \mathbf{w}) = (\mathbf{w}, \mathbf{v})$.
3. If $(\mathbf{v}, \mathbf{w}) = 0$ for all $\mathbf{w} \in V$ then $\mathbf{v} = \mathbf{0}$.

We observe that a scalar product is defined by specifying $(\mathbf{v}_i, \mathbf{v}_j)$ where $\{\mathbf{v}_1, \ldots, \mathbf{v}_n\}$ is a basis of V. In fact for any $\mathbf{v} = \alpha_1 \mathbf{v}_1, \ldots + \alpha_n \mathbf{v}_n$ and $\mathbf{w} = \beta_1 \mathbf{v}_1, \ldots + \beta_n \mathbf{v}_n$ we have

$$
(\mathbf{v}, \mathbf{w}) = \sum_{ij} \alpha_i \beta_j (\mathbf{v}_i, \mathbf{v}_j).
$$

Theorem 8.8: Let V be a vector space with a scalar product $(*, *)$ and let f be a functional over V then there exists $\mathbf{w} \in V$ so that

$$
f(\mathbf{v}) = (\mathbf{w}, \mathbf{v}).
$$

Proof: Let $\{\mathbf{v}_1, \ldots, \mathbf{v}_n\}$ be an orthonormal basis of V viz.

$$
(\mathbf{v}_i, \mathbf{v}_j) = \delta_{ij},
$$

and let $f(\mathbf{v}_i) = a_i$. We claim that the required $\mathbf{w}$ is given by

$$
\mathbf{w} = a_1 \mathbf{v}_1 + \cdots + a_n \mathbf{v}_n.
$$

In fact if $\mathbf{v} = (\alpha_1, \ldots, \alpha_n)$ then

$$
f(\mathbf{v}) = \alpha_1 f(\mathbf{v}_1) + \cdots + \alpha_n f(\mathbf{v}_n) = \alpha_1 a_1 + \cdots + \alpha_n a_n,
$$

and

$$
(\mathbf{w}, \mathbf{v}) = \left(\sum_i a_i \mathbf{v}_i, \sum_j \alpha_j \mathbf{v}_j \right) = \sum_{i,j} a_i \alpha_j (\mathbf{v}_i, \mathbf{v}_j) = \sum_{i,j} a_i \alpha_j \delta_{ij} = \alpha_1 a_1 + \ldots + \alpha_n a_n.
$$

Exercises

1. Find 1 forms ω in R^3 so that

   (a) $\omega \wedge d\omega = dx \wedge dy$

   (b) $\omega \wedge d\omega = dx \wedge dy \wedge dz$

2. Let $\omega = xdy - ydx$ and $\nu = x^2 dy + zdx + xydz$ be one forms on R^3. Compute

   (a) $d\omega$

   (b) $d\nu$

   (c) $\omega \wedge \nu$

   (d) $d(\omega \wedge \nu)$

3. Let $f(\mathbf{x})$, $g((\mathbf{x})$ be functions $R^n \to R^n$ with the the standard metric on R^n. Show that

$$df \wedge *dg = \sum_{i=1}^{n} \frac{\partial f}{\partial x_i} \frac{\partial g}{\partial x_i}.$$

4. Let V be a vector space of dimension n and T a linear transformation $T : V \to V$. Compute the determinant of $|\wedge^2 T|$

5. A matrix A is skew Hermitian if $A^\dagger + A = 0$ (where $A^\dagger = \bar{A}^T$ viz. transpose conjugate of A). Prove that $B = e^A$ is orthogonal i.e $BB^T = I$

6. A matrix A is skew symmetric if $A^T = -A$ show that the determinant of A is zero if the dimension of A is odd.

9

Integration on Manifolds in R^n

9.1 Integration in One-Dimension

We consider a one-dimensional manifold M in R^n, i.e., a curve. Assuming that M (or part of it) is represented parametrically by the mapping

$$\mathbf{f} = (f_1(t), \ldots, f_n(t))^T : (a, b) \subset R \rightarrow R^n,$$

then the length or "volume" of M is given by

$$V_1(M) = \int_a^b \left[\sqrt{\sum \left(\frac{df_i}{dt} \right)^2} \right] dt.$$

This formula can be rewritten in "vector form" as

$$V_1(M) = \int_a^b \sqrt{D\mathbf{f}^T\, D\mathbf{f}}\; dt,$$

where

$$D\mathbf{f}^T = \left(\frac{df_1}{dt}, \ldots, \frac{df_n}{dt} \right).$$

If we want to integrate a function $g = g(x_1, \ldots, x_n)$ on this manifold which is usually referred to as the line integral of g along the curve then

$$I(g) = \int_a^b g(\mathbf{f}) \sqrt{D\mathbf{f}^T D\mathbf{f}}\, dt.$$

It follows then that the volume element of this one-dimensional manifold is

$$dV = \sqrt{D\mathbf{f}^T D\mathbf{f}}\; dt.$$

Example 9.1: Let $r = r(\theta)$ be a the polar representation of a curve R^2 derive a formula for it length.

Solution: In Cartesian coordinates, the mapping $R \rightarrow R^2$ is

$$\theta \rightarrow (r(\theta) \cos \theta, r(\theta) \sin \theta),$$

DOI: 10.1201/9781003587422-9

hence

$$\frac{dx}{d\theta} = -r(\theta)\sin\theta + r'(\theta)\cos\theta, \quad \frac{dy}{d\theta} = r(\theta)\cos\theta + r'(\theta)\sin\theta.$$

Therefore

$$V = \int_{\theta_0}^{\theta_f} \sqrt{r^2 + (r')^2)}\, d\theta.$$

9.2 Volumes in R^3

It is well known that when one makes a change of from Cartesian coordinates (x, y, z) in R^3 to a new coordinate system $U = (u, v, w)$ the relationship between the volume elements in the two systems is given by

$$dV = dxdydz = \det(J)dudvdw,$$

where J is the Jacobian of the transformation viz.

$$J = \begin{pmatrix} \frac{\partial x}{\partial u} & \frac{\partial x}{\partial v} & \frac{\partial x}{\partial w} \\ \frac{\partial y}{\partial u} & \frac{\partial y}{\partial v} & \frac{\partial y}{\partial w} \\ \frac{\partial z}{\partial u} & \frac{\partial z}{\partial v} & \frac{\partial z}{\partial w} \end{pmatrix}. \tag{9.1}$$

Denoting by $\mathbf{F}$ the transformation

$$\mathbf{F} : (x, y, z) \to (x(u, v, w), y(u, v, w), z(u, v, w)).$$

we can rewrite $\det(J)^2$ as

$$\det(J)^2 = \text{Gram}\left(\frac{\partial \mathbf{F}}{\partial u}, \frac{\partial \mathbf{F}}{\partial v}, \frac{\partial \mathbf{F}}{\partial w}\right),$$

where

$$\text{Gram}(\mathbf{v}_1, \mathbf{v}_2, \mathbf{v}_3) = \det \begin{pmatrix} \mathbf{v}_1 \cdot \mathbf{v}_1 & \mathbf{v}_1 \cdot \mathbf{v}_2 & \mathbf{v}_1 \cdot \mathbf{v}_3 \\ \mathbf{v}_2 \cdot \mathbf{v}_1 & \mathbf{v}_2 \cdot \mathbf{v}_2 & \mathbf{v}_2 \cdot \mathbf{v}_3 \\ \mathbf{v}_3 \cdot \mathbf{v}_1 & \mathbf{v}_3 \cdot \mathbf{v}_2 & \mathbf{v}_3 \cdot \mathbf{v}_3 \end{pmatrix}. \tag{9.2}$$

Therefore we can rewrite the volume element as

$$dV = \sqrt{\text{Gram}\left(\frac{\partial \mathbf{F}}{\partial u}, \frac{\partial \mathbf{F}}{\partial v}, \frac{\partial \mathbf{F}}{\partial w}\right)}dudvdw.$$

The advantage of this formula is that the scalar product is well defined in R^n, and therefore, it can be generalized to the case where we consider a manifold M in R^n

Theorem 9.1: Let

$$\mathbf{F} : U \subset R^m \to R^n \ (t_1, \ldots, t_m) \to \mathbf{F}(t_1, \ldots, t_m),$$

be a manifold in R^n. The volume element in M is given by

$$dV_M = \sqrt{\left(\text{Gram}\left(\frac{\partial \mathbf{F}}{\partial t_1}, \ldots, \frac{\partial \mathbf{F}}{\partial t_m}\right)\right)} dt_1 \ldots dt_m.$$

This can be rewritten also as

$$dV_M = \sqrt{\det(J(\mathbf{F})^T J(\mathbf{F}))} dt_1 \ldots dt_m,$$

where $J(\mathbf{F})$ is the $n \times m$ Jacobian matrix

$$J(\mathbf{F})_{ij} = \frac{\partial F_i}{\partial t_j}.$$

We observe that for one-dimensional manifolds (i.e. $m = 1$) this formula reduces to the one derived in the previous section.

For a function defined on M, we have

$$\int_M f dV_M = \int_U f(\mathbf{F}(\mathbf{t}) dV_M.$$

Example 9.2: For two-dimensional manifolds in R^n the volume element is given by

$$dV = \sqrt{\left|\frac{\partial \mathbf{F}}{\partial t_1}\right|^2 \left|\frac{\partial \mathbf{F}}{\partial t_2}\right|^2 - \left(\frac{\partial \mathbf{F}}{\partial t_1} \cdot \frac{\partial \mathbf{F}}{\partial t_2}\right)^2} dt_1 dt_2.$$

We apply this formula to compute the surface area of the sphere R^3 with radius R. In spherical coordinates, the sphere is given by the mapping

$$\mathbf{F}(\theta, \phi) = (R \sin \phi \cos \theta, R \sin \phi \sin \theta, R \cos \phi).$$

Therefore

$$\frac{\partial \mathbf{F}}{\partial \phi} = (R \cos \phi \cos \theta, R \cos \phi \sin \theta, -R \sin \phi),$$

$$\frac{\partial \mathbf{F}}{\partial \theta} = (-R \sin \phi \sin \theta, R \sin \phi \cos \theta, 0).$$

Hence

$$dV = R^2 \sin \phi d\phi d\theta.$$

The area of the sphere is therefore

$$V = \int_0^{2\pi} \int_0^{\pi} R^2 \sin \phi d\phi d\theta = 4\pi R^2.$$

Example 9.3: A helicoid is defined by

$$\mathbf{F}(r,\theta) = (r\cos\theta, r\sin\theta, \theta),$$

where $0 \leq \theta \leq 2\pi$ and $0 \leq r \leq 1$. Compute its volume (viz surface area)

Solution:

$$\frac{\partial\mathbf{F}}{\partial\theta} = (-r\sin\theta, r\cos\theta, 1), \quad \frac{\partial\mathbf{F}}{\partial r} = (\cos\theta, \sin\theta, 0).$$

Hence the volume element is

$$dV = \sqrt{1+r^2}\,drd\theta,$$

The volume of the helicoid is therefore

$$V = \int_0^1 \int_0^{2\pi} \sqrt{1+r^2}\,drd\theta = 2\pi\int_0^1 \sqrt{1+r^2}\,dr = \pi[\sqrt{2}+\log(1+\sqrt{2})].$$

Example 9.4: A manifold M of dimension $(n-1)$ in R^n is called a hypermanifold. If such a hypermanifold is represented by

$$x_n = f(x_1,\ldots,x_{n-1}),$$

then its parametric representation is

$$\mathbf{F}(x_1,\ldots,x_{n-1}) = (x_1,\ldots,x_{n-1}, f(x_1,\ldots,x_{n-1})).$$

Therefore

$$\frac{\partial\mathbf{F}}{\partial x_1} = \left(1,\ldots,\frac{\partial f}{\partial x_1}\right), \quad \text{etc.}$$

A calculation then yields that

$$\text{Gram}\left(\frac{\partial\mathbf{F}}{\partial x_1},\ldots,\frac{\partial\mathbf{F}}{\partial x_{n-1}}\right) = 1 + \sum_{i=1}^{n-1}\left(\frac{\partial f}{\partial x_i}\right)^2,$$

and the volume element on the manifold is therefore

$$dV_M = \sqrt{1 + \sum_{i=1}^{n-1}\left(\frac{\partial f}{\partial x_i}\right)^2}\,dx_1\ldots dx_n.$$

Example 9.5: For the $(n-1)$ dimensional sphere S of radius R in R^n this formula yields

$$V(S_{n-1}) = \frac{2\pi^{n/2}}{\Gamma(n/2)}R^{n-1}.$$

For the volume (viz. surface area) of the two-dimensional sphere in R^3 this yields $V = 4\pi R^2$. Similarly for the three and four dimensional sphere in R^4 and R^5, respectively, we have

$$V(S_3) = 2\pi^2 R^3, \quad V(S_4) = \frac{8\pi^2}{3}R^4.$$

10

Integration on Manifolds

10.1 Simplicies in Euclidean Space

In spite of the fact that any manifold of dimension N can be embedded in Euclidean space of dimension $2n + 1$ (viz. E^{2n+1}), we shall pursue here the abstract integration theory on manifolds.

We start with the definition of simplicies in E^n

Definition 10.1

1. a 0-simplex is a point P_1.

2. A 1-simplex in a directed interval on a straight line which is characterized by it end points (P_1, P_2).

3. A 2-simplex is a triangle with ordered vertices (P_1, P_2, P_3).

In general, let $(P_1, \ldots, P_n)$ be a set of ordered points for which the vectors $(P_n - P_1), (P_{n-1} - P_1), \ldots, (P_2 - P_1)$ are linearly independent. An n-simplex is the "convex hull" of this set of points viz. the set of all points P which can be expressed as

$$P = t_1 P_1 + \cdots + t_n P_n, \quad t_1 + \cdots + t_n = 1.$$

An n-chain is defined as

$$c = \sum_{k=1}^{m} a_k s_k,$$

where each s_k is an n-simplex and a_k are constants.

The boundary of a simplex is defined as

$$\partial(P_1, \ldots, P_n) = \sum_{k=1}^{n} (-1)^{k-1}(P_1, \ldots, P_{k-1}, P_{k+1}, \ldots, P_n).$$

Thus the boundary of a simplex is a chain of $(n-1)$ simplicies.

   DOI: 10.1201/9781003587422-10

Example 10.1

1. $\partial(P_1, P_2) = (P_2) - (P_1)$
2. $\partial(P_1, P_2, P_3) = (P_2, P_3) - (P_1, P_3) + (P_1, P_2)$.

It is now easy to see that for any simplex s (and chain), "the boundary of the boundary" is 0. In fact if we apply the boundary operator to the n-simplex $(P_0, \ldots, P_n)$, then a typical "element" will be of the form $(P_0, \ldots, P_{i-1}, P_{i+1}, \ldots, P_{j-1}, P_{j+1}, \ldots, P_n)$. However, this element appears twice. (In the following, we assume without loss of generality that $j > i$) . It appears once as a result of applying the boundary operator to

$$(-1)^i (P_0, \ldots, P_{i-1}, P_{i+1}, \ldots, P_{j-1}, P_j, P_{j+1}, \ldots, P_n),$$

which yields

$$(-1)^{i+j-1}(P_0, \ldots, P_{i-1}, P_{i+1}, \ldots, P_{j-1}, P_{j+1}, \ldots, P_n).$$

It appears a second time by applying the boundary operator to

$$(-1)^j (P_0, \ldots, P_{i-1}, P_i, P_{i+1}, \ldots, P_{j-1}, P_{j+1}, \ldots, P_n),$$

which yields

$$(-1)^{i+j}(P_0, \ldots, P_{i-1}, P_{i+1}, \ldots, P_{j-1}, P_{j+1}, \ldots, P_n).$$

These two occurrences cancel each other, and therefore, we have in general that

$$\partial\partial s = 0.$$

Example 10.2

$$\partial(\partial(P_1, P_2, P_3, P_4))$$
$$= \partial((P_2, P_3, P_4) - (P_1, P_3, P_4) + (P_1, P_2, P_4) - (P_1, P_2, P_3)) \qquad (10.1)$$
$$= -([(P_3, P_4) - (P_2, P_4) + (P_2, P_3)] - [(P_3, P_4) - (P_1, P_4)) + (P_1, P_4)]$$
$$+ [(P_2, P4) - (P_1, P_4) + (P_1, P_2)] - (P_2, P_3) - (P_1, P_3) + (P_1, P_2)]) = 0.$$

Definition 10.2: The *standard simplex* in E^n is the simplex $\mathbf{S}^n = (S_0, \ldots, S_n)$ with $S_0 = (0, \ldots, 0)$ and for $k = 1, \ldots n$, $S_k = (0, \ldots, 1, \ldots, 0)$ where the 1 stands at the $k - th$ component of the vector.

Let $\omega = f(x_1, \ldots, x_n)dx_1 \wedge \ldots dx_n$ be an n-form which is defined on open set $U \subset E^n$ which contains $\mathbf{S}^n$ We define

$$\int_{\mathbf{S}^n} \omega = \int_{\mathbf{S}^n} f(x_1, \ldots, x_n)dx_1 \ldots dx_n.$$

10.2 Simplicies and Chains on Manifolds

A simplex on a manifold consists of a triplet (s^n, U, ψ) where s^n is a simplex in E^n, (i.e $s^n = (P_0, \ldots, P_n)$), $U \subset E^n$ and ψ is a (smooth but not necessarily one-to-one) mapping

$$\psi : U \to M.$$

We shall denote this simplex on M by τ.

For example, a 1-simplex on a manifold is a curve.

Another simplex on M, (r^n, W, ϕ) will be identified with the previous one if

$$\psi \left(\sum_{i=0}^{n} t_i P_i \right) = \phi \left(\sum_{i=0}^{n} t_i Q_i \right),$$

where $r^n = (Q_0, \ldots, Q_n)$.

If $\partial s^n = \sum_{i=0}^{n} \pm u_i$ we then define the boundary operator $\partial \tau$ as

$$\partial \tau = \sum_{i=0}^{n} \pm v_i,$$

where v_i are the images of the simplicies u_i under ϕ

Definition 10.3

- A **Cycle** is a chain C whose boundary is zero i.e. $\partial C = 0$.

- A chain B is a **Boundary** if it is a boundary of a chain i.e $B = \partial C$. It is obvious that every boundary is a cycle.

10.3 Integration of Forms on Manifolds

Let ω be a k-form on a manifold M and c a k-chain on M.

$$c = \sum_{i=0}^{n} a_i \tau_i,$$

where τ_i are k-simplecies on M and a_i are constants. To define

$$\int_c \omega,$$

we first let

$$\int_c \omega = \sum_{i=1}^n a_i \int_{\tau_i} \omega.$$

To define the last integral, we use the fact that there exists a triplet (S^k, U, ϕ) where S^k is the standard k-simplex in E^k and set

$$\int_{\tau_i} \omega = \int_{S^k} \phi^* \omega.$$

Since $\phi^* : M \to E^k$ the last integral is a regular k-iterated integral in E^k.

10.3.1   Stokes's Theorem

Let ω be a k-form on M and c a $k+1$ chain then

$$\int_{\partial c} \omega = \int_c d\omega.$$

Example 10.3: Let

$$\omega = (x^2 + 6y)dx + (-2x + y\cos(y^2))dy,$$

be a 1-form in R^2. Compute the integral of this form over the the boundary of the (triangular) simplex $s = ((0,0), (1,0), (0,1))$

Solution: By Stokes theorem we have

$$\int_{\partial s} \omega = \int_s d\omega,$$

but $d\omega = -8dx \wedge dy$ therefore

$$\int_{\partial s} \omega = -8 \int_s dx \wedge dy = -8 \int_0^1 \left(\int_0^{(1-x)} dy \right) dx = -4.$$

Example 10.4: Let f=f(x,y) satisfy Laplace equation on a 2-chain c in R^2, i.e. $\nabla^2 f = 0$ on c. Show that

$$\int_{\partial c} \left(\frac{\partial f}{\partial x} dy - \frac{\partial f}{\partial y} dx \right) = 0.$$

Solution: By Stokes theorem we have

$$\int_{\partial c} \left(\frac{\partial f}{\partial x} dy - \frac{\partial f}{\partial y} dx \right) = \int_c d\left(\frac{\partial f}{\partial x} dy - \frac{\partial f}{\partial y} dx \right) = \int_c \nabla^2 f dx \wedge dy = 0.$$

Example 10.5: Let C be the 1-chain

$$C : [0, 1] \to R^2 - \{0\}, \ C(t) = r(\cos 2\pi kt, \sin 2\pi kt),$$

where k is an integer. Show that C is not the boundary of a two chain in $R^2 - \{0\}$

Solution: Let ϕ be the angle function on C. Its differential $d\phi$ is well defined on C and we have

$$\int_C d\phi = 2\pi k.$$

Now suppose that there exists a 2-chain D in $R^2 - \{0\}$ so that $C = \partial D$ then by Stokes theorem we have

$$\int_C d\phi = \int_{\partial D} d\phi = \int_D d(d\phi) = 0.$$

which contradicts the previous result.

Example 10.6: Let ω be a 1-form on R^3

$$\omega = (x^2 + xy - z)dx \wedge dy + xdy \wedge dz - ydx \wedge dz,$$

and let $i : R^3 \to M$ be the inclusion map in R^3 to the unit disk in the x-y plane, i.e. $M = (x, y, z) \in R^3, x^2 + y^2 \le 1, z = 0$. Compute

$$\int_M i^*\omega.$$

Solution: Since $i^*\omega = (x^2 + xy)dx \wedge dy$ we have (using polar coordinates)

$$\int_M i^*\omega = \int_M (x^2 + xy)dxdy = \frac{\pi}{4}.$$

Example 10.7: Let ω be the following 1-form on $R^2 - \{0\}$

$$\omega = \frac{1}{2\pi(x^2 + y^2)}(xdy - ydx).$$

Show that ω is closed but not exact.

Solution:

$$d\omega = \frac{1}{2\pi(x^2 + y^2)^2}(-2xdx - 2ydy) \wedge (xdy - ydx) + \frac{1}{2\pi(x^2 + y^2)}2dx \wedge dy = 0.$$

To show that it is not exact we compute the integral of ω on the unit circle S^1 using polar coordinates

$$\int_{S^1} \omega = \frac{1}{2\pi} \int_0^{2\pi} (\cos^2 \theta + \sin^2 \theta) d\theta = 1.$$

However, if ω is exact then there should be a function f so that $df = \omega$. By Stokes theorem we should have then

$$\int_{S^1} \omega = \int_{S^1} df = \int_{\partial S^1} f = 0.$$

(The last integral is zero since the circle by itself is a boundary and the boundary of a boundary is zero). We obtained is a contradiction.

Example 10.8: Let ω be the following 2-form on $R^3 - \{0\}$

$$\omega = \frac{x dy \wedge dz + y dz \wedge dx + z dx \wedge dy}{(x^2 + y^2 + z^2)^{3/2}}.$$

Show that ω is closed but not exact.

Solution: To show that ω is closed we compute $d\omega$ and show that it is equal to 0. To show that it is not exact we compute its integral on the unit sphere in R^3 using spherical (geographical) coordinates

$$x = \cos \phi \cos \theta, \quad y = \cos \phi \sin \theta, \quad z = \sin \phi.$$

$\phi \in [-\pi/2, \pi/2], \theta \in [0, 2\pi]$. We obtain

$$\int_{S^2} \omega = \int_0^{2\pi} \int_{-\pi/2}^{\pi/2} -\cos \phi d\phi d\theta = -4\pi.$$

Now assume to the contrary that there exists a 1-form α so that $\omega = d\alpha$ then by Stokes theorem we have

$$\int_{S^2} \omega = \int_{S^2} d\alpha = \int_{\partial S^2} \alpha = 0.$$

This shows (as in the previous example) that ω is not exact.

10.4   Integral Theorems for Surfaces in R^3

In topology, the degree of a map is a numerical invariant that is related to a continuous mapping between two compact oriented manifolds of the same dimension. Intuitively, the degree represents the number of times that the domain manifold wraps around the range manifold under the mapping. It can proved (Brouwer,1911) that the degree is always an integer, but may be positive or negative depending on the orientations of the manifolds.

Let S be a closed surface in R^3 and let $\mathbf{e}_3$ be the outward normal to S. Now consider the mapping

$$\phi : S \to S^2, \quad \phi(\mathbf{x}) = \mathbf{e}_3(\mathbf{x}).$$

where S^2 is the unit sphere. As $\mathbf{x}$ varies on S, $\mathbf{e}_3$ wraps S^2 a whole number of times n. In particular if S is **convex** then $n = 1$. If

$$d\mathbf{e}_3 = \omega_1\mathbf{e}_1 + \omega_1\mathbf{e}_2, \tag{10.2}$$

then the area element of this map is

$$\omega_1 \wedge \omega_2 = K\sigma_1 \wedge \sigma_2, \tag{10.3}$$

where K is the Gaussian curvature of S. Therefore

$$\int_S K\sigma_1 \wedge \sigma_2 = \int_{S^2} \omega_1 \wedge \omega_2 = 4\pi n.$$

For closed convex surfaces $n = 1$, and this result says that the mean Gaussian curvature for these surfaces is always 4π.

For the rest of this section, we consider only closed convex surfaces. Two additional invariants for these surfaces are the total surface area A and the mean curvature M.

$$A = \int_S \sigma_1 \wedge \sigma_2, \quad M = \int_S H\sigma_1 \wedge \sigma_2.$$

We now want to find out how these invariants change if we translate every point $\mathbf{x}$ on S by a fixed distance α along $\mathbf{e}_3(x)$ i.e. the new surface $\bar{S}$ is

$$\mathbf{z} = \mathbf{x} + \alpha\mathbf{e}_3(\mathbf{x}),$$

and therefore using (10.2)

$$d\mathbf{z} = d\mathbf{x} + \alpha d\mathbf{e}_3 = (\sigma_1 + \alpha\omega_1)\mathbf{e}_1 + (\sigma_2 + \alpha\omega_2)\mathbf{e}_2.$$

This shows that the vectors $\mathbf{e}_1(\mathbf{x})$, $\mathbf{e}_2(\mathbf{x})$ span the tangent plane to $\bar{S}$ at $\mathbf{x}$ and $\mathbf{e}_3(x)$ is the normal at this point. Therefore

$$d\mathbf{z} = \rho_1\mathbf{e}_1 + \rho_2\mathbf{e}_2,$$

where

$$\rho_1 = \sigma_1 + \alpha\omega_1, \quad \rho_2 = \sigma_2 + \alpha\omega_2.$$

Therefore, the element of area on $\bar{S}$ is

$$\rho_1 \wedge \rho_2 = \sigma_1 \wedge \sigma_2 + \alpha(\sigma_1 \wedge \omega_2 - \sigma_2 \wedge \omega_1) + \alpha^2 \omega_1 \wedge \omega_2 = (1 + 2\alpha H + \alpha^2 K)\sigma_1 \wedge \sigma_2. \tag{10.4}$$

where we have used the definition of the mean curvature H. It follows then that the total area of $\bar{S}$ is

$$\bar{A} = \int_{\bar{S}} \rho_1 \wedge \rho_2 = \int_S (1 + 2\alpha H + \alpha^2 K)\sigma_1 \wedge \sigma_2 = A + 2\alpha M + 4\pi\alpha^2.$$

Using this relationship and integrating with to α on [0,alpha] we obtain for the new volume

$$\bar{V} = V + \alpha A + \alpha^2 M + \frac{4\pi\alpha^3}{3}.$$

Using the relationship

$$\omega_1 \wedge \omega_2 = K\sigma_1 \wedge \sigma_2 = \bar{K}\rho_1 \wedge \rho_2,$$

and (10.4) we obtain

$$\bar{K} = \frac{K}{(1 + 2\alpha H + \alpha^2 K)}.$$

Similarly by definition of the mean curvature, we have

$$\begin{aligned} 2\bar{H}\rho_1 \wedge \rho_2 &= \rho_1\omega_2 - \rho_2\omega_1 \\ &= (\sigma_1 + \alpha\omega_1) \wedge \omega_2 \\ &- (\sigma_2 + \alpha\omega_2) \wedge \omega_1 = 2H\sigma_1 \wedge \sigma_2 \\ &+ 2\alpha\omega_1 \wedge \omega_2 \\ &= 2(H + \alpha K)\sigma_1 \wedge \sigma_2. \end{aligned}$$

Using (5.3) we obtain

$$\bar{H} = \frac{H + \alpha K}{(1 + 2\alpha H + \alpha^2 K)}.$$

Finally

$$\bar{M} = \int_{\bar{S}} \bar{H}\rho_1 \wedge \rho_2 = \int_{\bar{S}} \frac{H + \alpha K}{1 + 2\alpha H + \alpha^2 K}\rho_1 \wedge \rho_2.$$

Using (5.3) we obtain

$$\bar{M} = \int_S (H + \alpha K)\sigma_1 \wedge \sigma_2 = M + 4\pi\alpha.$$

10.5   Relative Tensors and Integration on Non-Orientable Manifolds

10.5.1   Orientation

Definition 10.4: Let M be a (smooth) manifold. An atlas of M is oriented if for all charts (ϕ_α, U_α) and (ϕ_β, U_β) in the atlas with coordinate transformation $\psi_{\alpha\beta} : \phi_\beta \phi_\alpha^{-1}$ on $U_\alpha \cap U_\beta$ the determinant of $J(\psi_{\alpha\beta})$ is positive on its domain.

Definition 10.5: A manifold M is orientable if it possess an oriented atlas.

10.5.2   Relative Tensors

A relative tensor transforms under a change of coordinates as a regular tensor but in addition a "weight" is added as a coefficient.

For example if T_β^α is a mixed second rank tensor then its transformation under a change of coordinates is given by

$$\bar{T}_\beta^\alpha = \frac{\partial \bar{x}^\alpha}{\partial x^a} \frac{\partial x^b}{\partial \bar{x}^\beta} \det(J)^m T_b^a.$$

where m is an integer (positive or negative) and J is the Jacobian of the transformation.

$$J = \frac{\partial \bar{x}^i}{\partial x^j}.$$

We observe also that J^{-1} is given by

$$J^{-1} = \frac{\partial x^j}{\partial \bar{x}^i},$$

since

$$J J^{-1} = \frac{\partial \bar{x}^i}{\partial \bar{x}^j} = \delta_j^i,$$

When the "weight" $m = -1$ such relative tensors are referred to as *densities*.

How one can "generate" a physically meaningful relative tensor on a Riemannian manifold?

To this end, we consider the metric tensor g_{ij} which transforms as

$$\bar{g}_{ij} = \frac{\partial x^a}{\partial \bar{x}^i} \frac{\partial x^b}{\partial \bar{x}^j} g_{ab}.$$

This expression can re-interpreted in terms of matrix multiplication and hence when we take the determinant of these matrices we obtain

$$\bar{g} = (\det(J^{-1}))^2 g.$$

where g and $\bar{g}$ are the determinants of the metric tensors before and after the coordinate transformation. Observe that **that this result holds for orientable and non-orientable manifolds and the sign of g remains invariant under coordinate transformations.**

Assuming the both g and $\bar{g}$ are positive we can take the square root on both sides of this equation to obtain

$$\sqrt{\bar{g}} = \sqrt{g}|\det(J)^{-1}|. \tag{10.5}$$

If the manifold is orientable $\det(J)^{-1} > 0$ and $\sqrt{g}$ is a scalar (tensor) density i.e we can get rid of the absolute sign on $\det(J)^{-1}$ in (10.5).

As a side remark we note the if the determinant of g_{ij} is negative (as in Minkowiski space or General Relativity) we replace $\sqrt{g}$ by $\sqrt{-g}$ in the previous formulas.

Let **T** be a tensor (on an orientable manifold) then it now clear that $\mathbf{S} = \mathbf{T}\sqrt{g}$ is a density. For example if $\mathbf{T} = T_{ij}$ then

$$\bar{S}_{ij} = \bar{T}_{ij}\sqrt{\bar{g}} = \frac{\partial x^a}{\partial \bar{x}^i}\frac{\partial x^b}{\partial \bar{x}^j}T_{ab}\sqrt{g}\det(J)^{-1} = \det(J)^{-1}\frac{\partial x^a}{\partial \bar{x}^i}\frac{\partial x^b}{\partial \bar{x}^j}S_{ab}.$$

Remark 10.1: If ϕ is a mapping between two differential manifolds

$$\phi : M \to N,$$

then we can "lift" this mapping to the cotangent bundle (i.e the bundle of differential forms on these manifolds)

$$\phi^* : (TN)^* \to (TM)^*,$$

by the following definition

$$\phi^*(\omega)(X) = \phi_*(X)(\omega).$$

Where ω is a differential form in $(TN)^*$ and X is a vector field on M i.e. $X \in TM$.

In special case where ϕ is a (local) coordinate map

$$\phi : M \to R^n,$$

then

$$\phi^{-1} : R^n \to M,$$

and therefore

$$(\phi^{-1})^* : (TM)^* \to (TR^n)^*.$$

In other words, $(\phi^{-1})^*$ maps a differential form in $(TM)^*$ to it representation as a differential form defined on R^n. ∎

10.5.3   Integration

Consider an n-form ω on a compact oriented n-dimensional manifold and let the support of ω be contained in (ϕ_α, U_α). Furthermore let

$$(\phi_\alpha^{-1})^*(\omega) = f_\alpha(\mathbf{x})dx^1 \wedge \ldots \wedge dx^n,$$

then we define

$$\int_M \omega = \int_{\phi_\alpha(U_\alpha)} f_\alpha(\mathbf{x})dx^1 \ldots dx^n.$$

To show that this definition is independent of the chart being used suppose that the support of ω is contained also in (ϕ_β, U_β) which belongs to the same (orientable) atlas. On $U_\alpha \cap U_\beta$ we have

$$\phi_\alpha(m) = \psi_{\alpha\beta} \circ \phi_\beta(m), \quad \psi_{\alpha\beta} = \phi_\alpha \phi_\beta^{-1}, \quad m \in M.$$

Suppose now that

$$(\phi_\alpha^{-1})^*(\omega) = f_\alpha(\mathbf{x})dx^1 \wedge \ldots \wedge dx^n,$$

and

$$(\phi_\beta^{-1})^*(\omega) = g_\beta(\mathbf{x})dy^1 \wedge \ldots \wedge dy^n,$$

then

$$
\begin{aligned}
(\phi_\alpha^{-1})^*(\omega) &= (\phi_\beta^{-1} \circ \psi_{\alpha\beta}^{-1})^*(\omega)) \qquad\qquad\qquad (10.6)\\
&= (\psi_{\alpha\beta}^{-1})^* \circ (\phi_\beta^{-1})^*(\omega)\\
&= (\psi_{\beta\alpha})^* \circ (\phi_\beta^{-1})^*(\omega)\\
&= (\psi_{\beta\alpha}^*)(g_\beta(\mathbf{y})dy^1 \wedge \ldots \wedge dy^n)\\
&= (g_\beta \circ \psi_{\beta\alpha})(\mathbf{x}) \det(J(\psi_{\beta\alpha}))dx^1 \wedge \ldots \wedge dx^n.
\end{aligned}
$$

We observe that $\det(J(\psi_{\beta\alpha}))$ appears in this expression due to a change of coordinates. There is no absolute value sign on the determinant. Therefore

$$f_\alpha(\mathbf{x}) = g_\beta \circ \psi_{\beta\alpha}(\mathbf{x}) \det(J(\psi_{\beta\alpha})),$$

and by symmetry

$$g_\beta(\mathbf{y}) = f_\alpha \circ \psi_{\alpha\beta}(\mathbf{y}) \det(J(\psi_{\alpha\beta})).$$

It follows then that

$$\int_{U_\alpha} \omega = \int_{\phi_\alpha(U_{\alpha\beta})} f_\alpha(\mathbf{x}) dx^1 \dots dx^n$$

$$= \int_{\phi_\beta(U_{\alpha\beta})} f_\alpha(\psi_{\alpha\beta}(y)) \det(J(\psi_{\alpha\beta})) dy^1 \dots dy^n = \int_{U_\beta} \omega.$$

In general, the determinant of the Jacobian in the third integral should be taken in absolute value (otherwise a minus sign is possible). However, in this case, this does not make a difference since both charts belong to the same orientable atlas and $\det(J) > 0$. Thus, the definition of the integral of a form is independent of the coordinate chart being used.

On non-orientable manifolds it is impossible to define integration of tensorial quantities in a consistent way since there exist no orientable atlas, and therefore, the sign of $\det(J)$ might change under a coordinate transformation. To overcome this difficulty, we use densities.

Remark 10.2: To emphasize this point . The root of the problem is due to the fact that the transformation law for the volume element under a change of coordinates involves the **absolute value of the Jacobian determinant** while the transformation rule for n-form on an n-manifold involves the **Jacobian determinant** (without the absolute value). On non-orientable manifolds this may lead to ambiguity in the value of the integral in different coordinate systems.

Let dV be the volume element on a manifold M (**which is not necessarily orientable**). This volume element transforms under a coordinate transformation as

$$d\bar{V} = |J| dV.$$

It follows then from (3.3) that

$$\sqrt{\bar{g}} d\bar{V} = \sqrt{g} dV,$$

i.e $\sqrt{g} dV$ is a scalar tensor on the manifold. It is called the *invariant volume element on the manifold*. It follows then that

$$\int_M \sqrt{g} dV,$$

is an invariant. Similarly if ϕ is a scalar function on M then

$$\int_M \phi \sqrt{g} dV,$$

is also an invariant on M.

We note however that the integral of a tensor density S_{ab}

$$\int_M S_{ab} dV = \int_M T_{ab} \sqrt{g} dV,$$

has no well-defined transformation properties since it is not attached to a point on M, and there is no way to define its transformation coefficients.

11

Symmetry and Lie Groups

11.1 Definition of a Group

Let G be a set of elements with a well defined binary operation

$$G \times G \to G,$$

which has the following properties:

1. The operation is associative (ab)c=a(bc)

2. G has an element e so that for all $g \in G$, $eg = ge = g$

3. For every element $g \in G$ there exists $h \in G$ so that $gh = hg = e$ usually h is written as g^{-1}.

Note that in general the binary operation in G is not commutative viz. $g_1 g_2 \neq g_2 g_1$. When $g_1 g_2 = g_2 g_1$ for all elements of G we refer to G as a commutative group.

11.2 Introduction to Symmetry

Why symmetry is important?

Example 11.1: The Symmetries of the Equilateral Triangle

The following transformations leave this triangle unchanged

$$\begin{pmatrix} 1 & 2 & 3 \\ 1 & 2 & 3 \end{pmatrix}, \begin{pmatrix} 1 & 2 & 3 \\ 1 & 3 & 2 \end{pmatrix}, \begin{pmatrix} 1 & 2 & 3 \\ 3 & 2 & 1 \end{pmatrix}, \begin{pmatrix} 1 & 2 & 3 \\ 2 & 1 & 3 \end{pmatrix}, \tag{11.1}$$

$$\begin{pmatrix} 1 & 2 & 3 \\ 2 & 3 & 1 \end{pmatrix}, \begin{pmatrix} 1 & 2 & 3 \\ 3 & 2 & 1 \end{pmatrix}. \tag{11.2}$$

DOI: 10.1201/9781003587422-11

It follows then that the symmetry group of the equilateral triangle is the Group of permutations S_3.

Such discrete groups are important in solid state physics (crystals) and many other applications. However in many other (space-time problems, differential equations, etc.) groups with "infinitely many elements" have to be considered. For the rest of this chapter, we focus on this class of groups.

11.3   Lie Groups: ("Operational Definition")

A Lie group G is a group whose general element g can be expressed (continuously) in terms of n-parameters

$$g = g(\alpha_1 \ldots \alpha_n),$$

and the group multiplication in G varies continuously with $\boldsymbol{\alpha} = (\alpha_1, \ldots, \alpha_n)$.

The most important application and interest in Lie group is due to their action on R^n or Manifolds in general.

Example 11.2

1. Translations in $1 - dimension$

$$g(\alpha)(x) = x + \alpha.$$

2. Stretching in $1 - dimension$

$$g(\alpha)(x) = \alpha x \qquad \alpha \neq 0.$$

3. Rotations and reflections in $2 - dimenions$

A rotation in $2 - d$ (around the origin) is given by

$$T\begin{pmatrix} x \\ y \end{pmatrix} = \begin{pmatrix} \cos\theta & \sin\theta \\ -\sin\theta & \cos\theta \end{pmatrix}\begin{pmatrix} x \\ y \end{pmatrix}.$$

Examples of reflections are given by

$$\begin{array}{cc} x \to x & x \to -x \\ & \text{or} \\ y \to -y, & y \to y. \end{array}$$

Observe that reflections are not rotations by $\pm\pi$ (which take $x \to -x$, $y \to -y$). They are represented by

$$\begin{pmatrix} 1 & 0 \\ 0 & -1 \end{pmatrix} \text{ or } \begin{pmatrix} -1 & 0 \\ 0 & 1 \end{pmatrix}.$$

Geometrically both rotations and reflections preserve vector length $\| \mathbf{x} \|^2 = x^2 + y^2$

Definition 11.1: A $n \times n$ matrix O is orthogonal if

$$OO^T = I,$$

where O^T denote the transpose of O.

Theorem 11.1: The set $O(n)$ of $n \times n$ orthogonal matrices form a group that preserves $\| \mathbf{x} \|^2$

Proof: We first show that $O(n)$ is a Group. Let A, B be orthogonal matrices then

$$(AB)(AB)^T = ABB^T A^T = AA^T = I.$$

To show that it preserves $\| \mathbf{x} \|^2$ we compute:

$$\| O\mathbf{x} \|^2 = (O\mathbf{x}, O\mathbf{x}) = (O\mathbf{x})^T \cdot (O\mathbf{x}) = \mathbf{x}^T O^T O\mathbf{x} = \mathbf{x}^T \cdot \mathbf{x}.$$

This group is called $O(n)$. Observe that it contains both rotations and reflections.

To see this observe that

$$OO^T = I \Rightarrow \det(OO^T) = 1.$$

Therefore

$$(\det O)^2 = 1.$$

Hence

$$\det O = \pm 1.$$

The group has therefore two "sheets" one with $\det O = 1$ (pure rotations) and $\det O = -1$ which combine rotations with reflections. These two sheets are topologically separate since the mapping $O \to \det O$ is continuous and the point 0 has no pre-image in $O(n)$].

Sometimes one wants to restrict attention to rotations only and considers the subgroup:

$$SO(n) = \{O \in O(n); \det O = 1\}.$$

11.3.1   Spaces with Indefinite Metrics

We are used to the Euclidean space where

$$\| \mathbf{x} \|^2 = \mathbf{x} \cdot \mathbf{x} = \Sigma x_i^2 \equiv \mathbf{x}^T (I_n)\mathbf{x}.$$

where I_n is the n-dimensional unit matrix.

In relativity and various applications, we have to consider spaces with indefinite metrics, i.e.

$$I \rightarrow \begin{pmatrix} I_p & 0 \\ 0 & -I_{q,} \end{pmatrix}$$

where I_p, I_q are unit matrices of dimension p, q respectively.

In Minkowiski space-time a point is

$$\underline{x} = (\mathbf{x}, t) \quad \mathbf{x} \in R^3,$$

and

$$\underline{x} \cdot \underline{x} = -\mathbf{x} \cdot \mathbf{x} + t^2,$$

i.e.

$$\| \underline{x} \|^2 = (\mathbf{x}, t)^T \begin{pmatrix} -1 & & & O \\ & -1 & & \\ & & -1 & \\ O & & & 1 \end{pmatrix} \begin{pmatrix} x_1 \\ x_2 \\ x_3 \\ t \end{pmatrix}.$$

Let the metric be given by

$$I_{p,q} = \begin{pmatrix} -I_p & 0 \\ 0 & I_q \end{pmatrix}.$$

The group $O(p, q)$ is defined as

$$O(p, q) = \{O; O^T O = I_{p,q}\}.$$

The **Lorentz Group** is (isomorphic) to $O(3, 1)$. The **Poincare group** is $O(3, 1) \times T_4$ i.e the (semi-)direct product of the Lorentz group with the group of translations in four dimensions.

So far we considered Lie groups of transformations on R^n. We now consider these groups on C^n (vectors with complex entries).

Definition 11.2: A matrix U is unitary if

$$UU^\dagger = I \quad \text{where } \dagger = \text{transpose} + \text{complex conjugation}.$$

Theorem 11.2:

$$U(n) = \{U; U \text{ is } n \times n \text{ and unitary}\},$$

is a Lie group.

Remark 11.1: Observe that if a matrix O is $n \times n$ orthogonal matrix then O is unitary.

If we define

$$\| \mathbf{x} \|^2 = \mathbf{x}^\dagger \cdot \mathbf{x} = \sum_{i=1}^{n} x_i^* x_i.$$

Then the group $U(n)$ is the set of all transformations that preserve this metric. A subgroup of $U(n)$ is

$$SU(n) = \{U \in U(n), \det U = 1\}.$$

11.3.1.1 Other Important Examples

$GL(n, C)$ is group of all linear transformations

$$T : C^n \to C^n, \quad \det T \neq 0.$$

Similarly

$$SL(n, C) = \{T \in GL(n, C), \quad \det T = 1\}.$$

Sometimes one considers also $GL(n, R)$ and $SL(n, R)$.

11.3.1.2 Symplectic Groups

Definition 11.3: The symplectic group $Sp(n, R)$ is defined as

$$Sp(n, R) = \left\{ A \in M(2n, 2n, R) \mid A^T J A = J \right\},$$

where $M(2n, 2n, R)$ is the set of all $2n \times 2n$ real matrices and

$$J = \begin{pmatrix} 0 & -I_n \\ I_n & 0 \end{pmatrix},$$

and I_n is the n-dimensional unit matrix.

11.4   Lie Algebras

We want to generalize the concept of a "basis of a vector space" to Lie group. However in general the group operation here is non-commutative.

Definition 11.4: Let A be an "operator" ($\equiv$ matrix or differential operator) we define

$$e^{\alpha A} = \sum_{n=0}^{\infty} \frac{\alpha^n}{n!} A^n.$$

Example 11.3: Let $A = \frac{d}{dx}$ and $f(x)$ analytic viz. $f(x) \in C^\infty(R)$ and its Taylor expansion converges to f.

$$e^{\alpha A} f(x) = \sum_{n=0}^{\infty} \frac{\alpha^n}{n!} \frac{d^n}{dx^n} f(x) = f(x + \alpha).$$

We see from this example that the action of $e^{\alpha \frac{d}{dx}}$ on any analytic function $f(x)$ is equivalent to translation of x by α.

Example 11.4:

$$A = \begin{pmatrix} 0 & -i \\ i & 0 \end{pmatrix}.$$

We observe that $A^2 = I$ and therefore we have

$$e^{i\alpha A} = I + i\alpha A - \frac{\alpha^2}{2!} I - \frac{i\alpha^3}{3!} A + + \frac{\alpha^4}{4!} I + \frac{i\alpha^5}{5!} A + \cdots.$$

Hence

$$e^{i\alpha A} = \cos \alpha I + i \sin \alpha A,$$

i.e.

$$e^{i\alpha A} = \begin{pmatrix} \cos \alpha & \sin \alpha \\ -\sin \alpha & \cos\alpha \end{pmatrix}.$$

In other words, $e^{i\alpha A}$ represents a rotation by angle α in R^2.

Example 11.5: Let $D = x\frac{d}{dx}$

We claim that

$$e^{\alpha D} f(x) = f(e^\alpha x).$$

To prove this result, we first note that for $m, n > 0$ we have

$$D^n x^m = m^n x^m.$$

(This can be proved by induction). Using the Taylor expansion of $f(x)$ around zero we then have

$$\begin{aligned}
e^{\alpha D} f(x) &= e^{\alpha D} \sum_{m=0}^{\infty} \frac{f^m(0)}{m!} x^m = \sum_{m=0}^{\infty} \frac{f^m(0)}{m!} e^{\alpha D} x^m \\
&= \sum_{m=0}^{\infty} \frac{f^m(0)}{m!} \sum_{n=0}^{\infty} \frac{\alpha^n}{n!} m^n x^m \\
&= \sum_{m=0}^{\infty} \frac{f^m(0)}{m!} x^m \sum_{n=0}^{\infty} \frac{(\alpha m)^n}{n!} = \sum_{m=0}^{\infty} \frac{f^m(0)}{m!} x^m e^{\alpha m} \\
&= \sum_{m=0}^{\infty} \frac{f^m(0)}{m!} (e^\alpha x)^m = f(e^\alpha x).
\end{aligned}$$

(11.3)

It follows then that the operator D is the generator of dilatations (or "stretching") in one dimension.

Definition 11.5: A Lie algebra is a vector space of operators (with respect to addition and multiplication by a scalar) with generators L_i, $i = 1, \ldots, n$ which satisfy

$$[L_i, L_j] = L_i L_j - L_j L_i = \sum_{k=1}^{n} c_{ij}^k L_k,$$

where c_{ij}^k are constants (which are referred to as the structure constants of the algebra) and

$$[L_i, [L_j, L_k]] + [L_j, [L_k, L_i]] + [L_k, [L_i, L_j]] = 0, \quad \text{(Jacobi identity)}.$$

In other words a Lie algebra is a vector space of operators which is closed under the commutator operation and satisfies the Jacobi identity.

If we can find a Lie algebra $\mathcal{L} = \{L_1 \ldots L_n\}$ so that for every g in the Lie group G there exists $L \in \mathcal{L}$ which satisfies

$$e^L = g.$$

we then say that $\mathcal{L}$ is the Lie algebra of G.

Example 11.6: The operators

$$J_1 = \left(z \frac{\partial}{\partial y} - y \frac{\partial}{\partial z} \right)$$

$$J_2 = \left(x \frac{\partial}{\partial z} - z \frac{\partial}{\partial x} \right)$$

$$J_3 = \left(y \frac{\partial}{\partial x} - x \frac{\partial}{\partial y} \right),$$

form a Lie algebra

$$[J_i, J_j] = \epsilon_{ijk} J_k \quad \text{(summation over } k\text{)},$$

$$\epsilon_{ijk} = \begin{cases} 1 & (i, j, k) \text{ is even permutation of } (1, 2, 3) \\ -1 & (i, j, k) \text{ is odd permutation of } (1, 2, 3) \\ 0 & \text{otherwise.} \end{cases}$$

This is the Lie algebra of $SO(3)$.

11.5   Manifolds and Lie Groups

Definition 11.6: A Lie group is a group G which is also a manifold and for which the mapping

$$G \times G \to G, \quad (g_1, g_2) \to g_1 g_2,$$

is differentiable.

Example 11.7:

1. R^n is an additive group and a manifold. the mapping $(x, y) \to x + y$ is differentiable. Hence R^n is a Lie group.

2. $C^* = C - 0$ that is the complex numbers without 0 is a Lie group.

The natural global chart on C^* is $z = a + ib \to (a, b) \in R^2$ is obviously differentiable and hence C^* is a manifold. C^* is also a group with respect to complex number multiplication. The mapping $T : C^* \times C^* \to C^*$ defined by $T(z_1, z_2) = z_1 z_2$ is differentiable. Hence C^* is a Lie group.

3. We already showed that the unit circle S^1 is a manifold. If we represent the points of S^1 by $e^{i\theta}$ then it is obviously a group with respect to multiplication. The mapping $T(e^{i\theta}, e^{i\phi}) = e^{i(\theta+\phi)}$ is differentiable.

4. $GL(n, R)$ the set of all $n \times n$ nonsingular matrices is a Lie group.

For each element $h \in G$ we define the left translation

$$L_h : G \to G, \quad g \to hg,$$

since the group multiplication is differentiable the mapping L_h is differentiable. Since $L_{h^{-1}}$ is also differentiable L_h is a diffeomorphism of G onto itself.

Right translations can be defined in a similar way $R_h : g \to gh$.

Theorem 11.3: Let G be a Lie group then $\phi : G \to G$ defined by $g \to g^{-1}$ is a diffeomorphism.

Definition 11.7: Let G be a Lie group and h_* the differential of L_h (i.e. $h_* : TG \to TG$) we shall say that a vector field X on G is left invariant if

$$X(hg) = h_*(X(g)),$$

that is the vector representing X at the point hg is equal to its value at $X(g)$ when this vector is is mapped by h_*. We observe that a left invariant vector field is determined by its value at the unit element e since by the above definition

$$X(h) = h_*(X(e)).$$

From this observation we infer that the (Lie) bracket operation on the set of left invariant vector fields on G form a Lie algebra $\mathcal{L}G$. In fact if $u, v \in T_e G$ and X, Y are the corresponding left invariant vector fields $X(e) = u$, $Y(e) = v$ then

$$[X, Y](e) = [u, v].$$

(Since a left invariant vector field is determined by its value at e and the vector fields on the left and right have the same value at this point).

The dimension of $\mathcal{L}G$ is equal to the dimension of G and it is isomorphic to $T_e G$.

It follows then that if X_i, X_j are left invariant vector fields then

$$[X_i, X_j] = \sum_k C_{ij}^k X_k.$$

The constants C_{ij}^k are called the structure constants of the algebra.

Example 11.8: Let G consist of the set

$$G : \left\{ g \middle| g = \begin{pmatrix} z & 0 & y \\ 0 & z & x \\ 0 & 0 & 1 \end{pmatrix}, \ x, y, z \in R \ z \neq 0 \right\}.$$

It is easy to show that G is a group with respect to matrix multiplication with

$$g^{-1} = \begin{pmatrix} 1/z & 0 & y/z \\ 0 & 1/z & x/z \\ 0 & 0 & 1 \end{pmatrix}.$$

(In fact G is a subgroup and submanifold of $GL(3, R)$).

A global chart on G is given by $g \to (x, y, z)$. (To simplify the notation we write this as $g(x, y, z)$). The tangent plane at any point of G is the vector space generated by

$$\frac{\partial}{\partial x}, \ \frac{\partial}{\partial y}, \ \frac{\partial}{\partial z}.$$

To determine the left invariant vector fields on TG we use the formula

$$X(h) = h_*(X(e)).$$

If $h = h(\alpha, \beta, \gamma)$ then the left translation generated by h is

$$h(g) = hg = \begin{pmatrix} \gamma z & 0 & \gamma y + \beta \\ 0 & \gamma z & \gamma x + \alpha \\ 0 & 0 & 1 \end{pmatrix} = \bar{g}(\gamma x + \alpha, \gamma y + \beta, \gamma z).$$

Hence

$$h_* \left(\frac{\partial}{\partial x} \right)_e = \gamma \left(\frac{\partial}{\partial x} \right)_h,$$

$$h_* \left(\frac{\partial}{\partial y} \right)_e = \gamma \left(\frac{\partial}{\partial y} \right)_h,$$

$$h_* \left(\frac{\partial}{\partial z} \right)_e = \gamma \left(\frac{\partial}{\partial z} \right)_h.$$

Since γ is the z coordinate of h it follows that the left invariant vector fields on G at the point $g(x, y, z)$ are

$$z \left(\frac{\partial}{\partial x} \right), \; z \left(\frac{\partial}{\partial y} \right), \; z \left(\frac{\partial}{\partial z} \right).$$

It is easy to show that these three operators form a Lie algebra.

Example 11.9: For $GL(n, R)$, $\mathcal{L}GL(n, R)$ is formed by

$$X_{ij} = \sum x_{ik} \frac{\partial}{\partial x_{kj}},$$

and the commutation relations are given by

$$[X_{ij}, X_{kl}] = \delta_{kj} X_{il} - \delta_{il} X_{kj}.$$

11.5.1 Lie Transformation Groups

A Lie group G acts as a transformation group on a manifold M if there exists a surjective differentiable map

$$\psi : G \times M \to M,$$

which satisfies

$$\psi(g_1, \psi(g_2, m)) = \psi(g_1 g_2, m),$$

where $g_1, g_2 \in G$ and $m \in M$.

Example 11.10: the group $GL(n, R)$ acts as a Lie transformation group on R^n with

$$\psi(A, \mathbf{v}) = A\mathbf{v}, \quad A \in GL(n, R), \quad \mathbf{v} \in R^n.$$

Example 11.11: A Lie group G acts as Lie transformation group on itself using the group mapping $G \times G \to G$.

Definition 11.8: The orbit of $m \in M$ under G is the set $\{h \in M | h = \psi(g, m) \text{ for some } g \in G\}$

The Lie derivative of of a scalar function $f : M \to R$ respect to a vector field X on M is defined as

$$\mathcal{L}_X(f)(p) = X(f)(p).$$

The Lie derivative of of a vector field Y with respect to a vector field X is defined as

$$\mathcal{L}_X(Y) = [X, Y].$$

These definitions can be generalized to tensorial vector fields and forms.

11.6 The Exponential Map

The exponential map relates the Lie algebra back to the Lie group.

Theorem 11.4: If $\mathbf{L} \in (TG)_e$ and $a, b \in R$ then

$$e^{(a+b)\mathbf{L}} = e^{a\mathbf{L}} e^{b\mathbf{L}}.$$

Theorem 11.5: For any one-parameter subgroup H of a connected Lie group G there exists $\mathbf{L} \in (TG)_e$ so that for all $g \in H$ we have $g = e^{\alpha \mathbf{L}}$.

Let G be a Lie group. The Lie subgroup of G that is generated by an element L of the associated Lie algebra is given by exponentiation viz. $e^{\alpha L}$, $\alpha \in R$ is an element of G. When L is represented by a matrix A the computation of g requires the evaluation of $e^{\alpha A}$. In this section we describe the algorithm needed to accomplish this objective.

The definition of a matrix exponential uses the Taylor expansion of e^x around zero with x being replaced by A.

Definition 11.9: Let A be $n \times n$ matrix.

$$e^{\alpha A} = \sum_0^\infty \frac{(\alpha A)^n}{n!}. \tag{11.4}$$

11.6.1 Some Basic Properties

1. If A, B are $n \times n$ matrices and A,B commute viz $AB = BA$ then

   $$e^{A+B} = e^A e^B = e^B e^A,$$

   this follow from the fact that when A,B commute their multiplication "behaves" as those for real numbers e.g.,

   $$(A+B)^2 = (A+B)(A+B) = A^2 + AB + BA + B^2 = A^2 + 2AB + B^2.$$

2. A matrix O with real elements is orthogonal if

   $$OO^T = I.$$

Thus the transpose of O equals its inverse.

3. A matrix H is Hermitian if

$$H = H^\dagger,$$

where $H^\dagger$ stands for the complex conjugate transpose of H. (viz. Take the complex conjugate of each element of H and then apply the transpose operation. If H is Hermitian then the result should be equal to H.

4. A matrix U is unitary if

$$UU^\dagger = I.$$

Thus a unitary matrix is one whose inverse equals it complex-conjugate transpose of itself. Thus orthogonal matrices are unitary.

5. **Theorem 11.6**: If the matrix H is Hermitian then $U = e^{iH}$ is unitary

Proof: By definition of the matrix exponential we have

$$\exp^{iH} = \sum_0^\infty \frac{(iH)^n}{n!}.$$

Hence

$$(\exp^{iH})^\dagger = \sum_0^\infty \frac{(-i)^n * (H^\dagger)^n}{n!} = \sum_0^\infty \frac{(-1)^n (i * H)^n}{n!} = \exp^{-iH},$$

since $e^{iH}e^{-iH} = I$ this proves that $\exp^{iH}$ is unitary.

11.6.2 Computation of the Exponential

Let A be $n \times n$ matrix to compute the exponential of A we use the following theorem;

Theorem 11.7: Cayley-Hamilton Let $f(x)$ be a smooth function and A, $n \times n$ matrix. There exists a polynomial $r(x)$ of degree $(n-1)$ so that $f(A) = r(A)$ (viz. x is being replaced by the matrix A in these functions). Thus if $r(x) = a_0 + a_1 x + \cdots + a_{n-1} x^{n-1}$ then

$$f(A) = a_0 I + a_1 A + \cdots + a_{n-1} A^{n-1},$$

where I is the unit matrix. To compute the coefficients of $r(x)$ we have the following:

Theorem 11.8: Assume the matrix A has n distinct eigenvalues $\lambda_1, \ldots, \lambda_n$ then

$$f(\lambda_i) = r(\lambda_i).$$

This yields the necessary n equations to compute the coefficients of $r(x)$

Example 11.12: Let

$$A = \begin{pmatrix} 1 & 2 \\ 4 & 3 \end{pmatrix}, \tag{11.5}$$

compute e^A.

Solution: For this example $f(x) = e^x$. The characteristic polynomial of A is

$$p(\lambda) = \begin{vmatrix} 1-\lambda & 2 \\ 4 & 3-\lambda \end{vmatrix} = (1-\lambda)(3-\lambda) - 8 = \lambda^2 - 4\lambda - 5. \tag{11.6}$$

Hence the eigenvalues of A are $\lambda_1 = 5$, $\lambda_2 = -1$. Since A is a 2×2 matrix $r(x) = a_0 + a_1 x$ and we must have

$$e^{\lambda_i} = a_0 + a_1\lambda_i, \; i = 1, 2.$$

Solving this system of equations we obtain

$$a_0 = \frac{e^5 + 5e^{-1}}{6}, \; a_1 = \frac{e^5 - e^{-1}}{6}.$$

Hence denoting the unit matrix in two dimension by I we have

$$e^A = a_0 I + a_1 A = \begin{pmatrix} \frac{e^5+2e^{-1}}{3} & \frac{e^5-e^{-1}}{3} \\ \frac{2e^5-2e^{-1}}{3} & \frac{2e^5+e^{-1}}{3} \end{pmatrix}. \tag{11.7}$$

If the matrix A has some eigenvalues with multiplicity greater that one the number of equations for the coefficients of $r(x)$ (obtained by Theorem 11.7) will bo loss than n. To deal with these situations we have the following extension of Theorem 11.7:

Theorem 11.9: If λ is an eigenvalue of multiplicity k of the matrix A then
$$f^j(\lambda) = r^j(\lambda), \; j = 0, \ldots (k-1),$$

where f^j and r^j are the derivatives of order j of these functions with respect to x.

Example 11.13: Let

$$A = \begin{pmatrix} -1 & -1 \\ 1 & -3 \end{pmatrix}, \tag{11.8}$$

compute e^{At}.

Solution: In this case, the matrix to exponentiate is $B = At$ and the function $f(x) = e^x$. The eigenvalues of the matrix B are the roots of the characteristic polynomial

$$p(\lambda) = \begin{vmatrix} -t-\lambda & -t \\ t & -3t-\lambda \end{vmatrix} = (-t-\lambda)(-3t-\lambda)+t^2 = \lambda^2 + 4\lambda t + 4t^2 \quad (11.9)$$

Hence, the matrix B has one eigenvalue $\lambda = -2t$ with multiplicity 2. Since B is 2×2 matrix $r(x) = a_0 + a_1 x$ and by Theorem 11.8 we have

$$f(\lambda) = r(\lambda), \ f'(\lambda) = r'(\lambda).$$

Therefore

$$e^{-2t} = a_0 + a_1(-2t), \ e^{-2t} = a_1.$$

Hence

$$a_0 = e^{-2t}(1+2t), \ a_1 = e^{-2t}.$$

It follows then that

$$e^B = e^{At} = a_0 I + a_1(At) = e^{-2t}\begin{pmatrix} 1+t & -t \\ t & 1-t \end{pmatrix}. \quad (11.10)$$

11.6.3 Matrix Representation of some Lie Algebras

Here we provide examples for the matrix represention of some Lie algebras.

1. The Lie algebra SO(3) is represented by the following 2×2 matrices which are usually referred to as Pauli matrices.

$$J_1 = \begin{pmatrix} 0 & 1 \\ 1 & 0 \end{pmatrix}, \ J_2 = \begin{pmatrix} 0 & -i \\ i & 0 \end{pmatrix}, \ J_3 = \begin{pmatrix} 1 & 0 \\ 0 & -1 \end{pmatrix}. \quad (11.11)$$

the commutation relations for these matrices are

$$[J_1, J_2] = iJ_3, \ [J_2, J_3] = iJ_1, \ [J_3, J_1] = iJ_2.$$

2. The Lie algebra SU(3) is represented by the following eight 3×3 matrices which are usually referred to as Gell-Mann matrices.

$$\lambda_1 = \begin{pmatrix} 0 & 1 & 0 \\ 1 & 0 & 0 \\ 0 & 0 & 0 \end{pmatrix}, \ \lambda_2 = \begin{pmatrix} 0 & -i & 0 \\ i & 0 & 0 \\ 0 & 0 & 0 \end{pmatrix}, \quad (11.12)$$

$$\lambda_3 = \begin{pmatrix} 1 & 0 & 0 \\ 0 & -1 & 0 \\ 0 & 0 & 0 \end{pmatrix}, \ \lambda_4 = \begin{pmatrix} 0 & 0 & 1 \\ 0 & 0 & 0 \\ 1 & 0 & 0 \end{pmatrix} \quad (11.13)$$

$$\lambda_5 = \begin{pmatrix} 0 & 0 & -i \\ 0 & 0 & 0 \\ i & 0 & 0 \end{pmatrix}, \quad \lambda_6 = \begin{pmatrix} 0 & 0 & 0 \\ 0 & 0 & 1 \\ 0 & 1 & 0 \end{pmatrix}, \quad (11.14)$$

$$\lambda_7 = \begin{pmatrix} 0 & 0 & 0 \\ 0 & 0 & -i \\ 0 & i & 0 \end{pmatrix}, \quad \lambda_8 = \frac{1}{\sqrt{3}}\begin{pmatrix} 1 & 0 & 1 \\ 0 & 1 & 0 \\ 0 & 0 & -2 \end{pmatrix}. \quad (11.15)$$

The commutation relations between these generators are

$$[\lambda_m, \lambda_n] = 2if_{mnk}\lambda_k,$$

where f_{mnk} are antisymmetric under the permutation of any pair of indices and a tabulation of their nonzero values is as follows:

$$f_{123} = 1, \ f_{156} = -\frac{1}{2}, \ f_{246} = \frac{1}{2},$$

$$f_{257} = \frac{1}{2}, \ f_{345} = \frac{1}{2}, \ f_{367} = -\frac{1}{2},$$

$$f_{458} = \frac{\sqrt{3}}{2}, \ f_{678} = \frac{\sqrt{3}}{2}.$$

3. The Lie algebra of the Lorenz group SO(3,1) consists of six generators. Three of these J_1, J_2 and J_3 represent rotations and the other three K_1, K_2 and K_3 represent boosts. A matrix representation of these generators is as follows;

$$J_1 = i\begin{pmatrix} 0 & 0 & 0 & 0 \\ 0 & 0 & 0 & 0 \\ 0 & 0 & 0 & -1 \\ 0 & 0 & 1 & 0 \end{pmatrix}, \quad J_2 = i\begin{pmatrix} 0 & 0 & 0 & 0 \\ 0 & 0 & 0 & 1 \\ 0 & 0 & 0 & 0 \\ 0 & -1 & 0 & 0 \end{pmatrix},$$

$$J_3 = i\begin{pmatrix} 0 & 0 & 0 & 0 \\ 0 & 0 & -1 & 0 \\ 0 & 1 & 0 & 0 \\ 0 & 0 & 0 & 0 \end{pmatrix} \quad (11.16)$$

$$K_1 = i\begin{pmatrix} 0 & 1 & 0 & 0 \\ 1 & 0 & 0 & 0 \\ 0 & 0 & 0 & 0 \\ 0 & 0 & 0 & 0 \end{pmatrix}, \quad K_2 = i\begin{pmatrix} 0 & 0 & 1 & 0 \\ 0 & 0 & 0 & 0 \\ 1 & 0 & 0 & 0 \\ 0 & 0 & 0 & 0 \end{pmatrix},$$

$$K_3 = i\begin{pmatrix} 0 & 0 & 0 & 1 \\ 0 & 0 & 0 & 0 \\ 0 & 0 & 0 & 0 \\ 1 & 0 & 0 & 0 \end{pmatrix}. \quad (11.17)$$

The commutation relations between these generators are

$$[J_i, J_j] = i\epsilon_{ijk}J_k, \ [J_i, K_j] = i\epsilon_{ijk}K_k, \ [K_i, K_j] = -i\epsilon_{ijk}J_k.$$

11.7　Applications

A. Similarity solutions

Consider the heat equation

$$\frac{1}{k}\frac{\partial u}{\partial t} = \nabla^2 u \qquad (\dim = 1, 2, 3),$$

assuming $u = u(r,t)$ it is straightforward to see that this equation remains invariant (unchanged) under the transformations

$$\bar{r} = \alpha r, \quad \bar{t} = \alpha^2 t \quad \bar{u} = \beta u + \gamma,$$

where α, β, γ are constants (a three-parameter Lie group). For any positive n this three-parameter group contains the following one-parameter subgroup

$$\bar{r} = \alpha r, \quad \bar{t} = \alpha^2 t \quad \bar{u} = \alpha^n u .$$

A similarity solution is a solution that is invariant under such a one-parameter group. In fact if we define

$$\eta = r^2/t \quad u = t^{n/2} f(\eta),$$

and substitute in the heat equation we obtain

$$4k\eta f'' + \left(\eta + \frac{n}{2}\right) f' - \frac{m}{2} f = 0,$$

which is a hypergeometric equation (after a substitution $x = -\eta$)

　　Thus by a proper transformation we were able to reduce the PDE to an ODE.

　　Example 11.14: As another example for the application of similarity transformations we consider a nonlinear heat equation where the diffusion coefficient depends on u.

$$\frac{\partial u}{\partial t} = \frac{\partial}{\partial x}\left(D(u)\frac{\partial u}{\partial x}\right). \qquad (11.18)$$

To find a similarity solution we attempt to introduce a new variable $\eta = x^\alpha t^\beta$ where α and β are constants to be determined latter and attempt to find a solution of (11.18) which is dependent only on η viz. $u = u(\eta)$. For such a function, we have (primes denote differentiation with respect to η).

$$\frac{\partial u}{\partial t} = \beta x^\alpha t^{\beta-1} u',$$

$$\frac{\partial u}{\partial x} = \alpha x^{\alpha-1} t^{\beta} u',$$

and

$$\frac{\partial}{\partial x}\left(D(u)\frac{\partial u}{\partial x}\right) = \alpha(\alpha-1)x^{\alpha-2}t^{\beta}D(u)u' + \alpha^2 x^{2\alpha-2}t^{2\beta}\left[D(u)u'\right]'.$$

Substituting these results in (11.18) and dividing by $x^{\alpha-2}t^{\beta}$ we obtain

$$\beta\frac{x^2}{t}u'(\eta) = \alpha(\alpha-1)D(u)u' + \alpha^2\eta\left[D(u)u'\right]'.$$

We see that if we choose $\alpha = 1$ and $\beta = -1/2$, i.e.

$$\eta = \frac{x}{t^{1/2}},$$

then the explicit dependence of this equation on x, t separately disappears. The equation simplifies considerably and we have

$$[D(u)u']' + \frac{1}{2}\eta u' = 0. \tag{11.19}$$

i.e., the partial differential equation has been reduced to an ODE. This transformation is due to Boltzmann.

Exercises

1. Compute the Lie group elements of $SO(3)$ that are represented by $e^{\alpha J_1}$, $e^{\alpha J_2}$ and $e^{\alpha J_3}$.

2. Compute the Lie group elements of $SO(3,1)$ that are represented by $e^{\alpha J_1}$, $e^{\alpha K_1}$ and $e^{\alpha K_3}$

3. Compute the Lie group elements of $SU(3)$ that are represented by $e^{\alpha\lambda_6}$, $e^{\alpha\lambda_7}$ and $e^{\alpha\lambda_8}$

4. Prove that if a matrix A is antihermitian (viz. $H^{\dagger} = -H$) then e^A is unitary.

5. Show that the following set of differential operators satisfy the commutation relations for the Lie algebra of the Lorentz group $SO(3,1)$.

$$J_1 = -i\left(y\frac{\partial}{\partial z} - z\frac{\partial}{\partial y}\right), \quad J_2 = -i\left(z\frac{\partial}{\partial x} - x\frac{\partial}{\partial z}\right),$$

$$J_3 = -i\left(x\frac{\partial}{\partial y} - y\frac{\partial}{\partial x}\right)$$

$$K_1 = -i\left(t\frac{\partial}{\partial x} + x\frac{\partial}{\partial t}\right), \; K_2 = -i\left(t\frac{\partial}{\partial y} + y\frac{\partial}{\partial t}\right),$$

$$K_3 = -i\left(t\frac{\partial}{\partial z} + z\frac{\partial}{\partial t}\right). \tag{11.20}$$

6. compute e^{At} where A is the following matrix

$$A = \begin{pmatrix} 2 & 2 & -4 \\ 2 & -1 & -2 \\ 4 & 2 & -6 \end{pmatrix}, \tag{11.21}$$

Hint: One of the eigenvalues of the matrix A is -1.

Index

For Product Safety Concerns and Information please contact our EU
representative GPSR@taylorandfrancis.com
Taylor & Francis Verlag GmbH, Kaufingerstraße 24, 80331 München, Germany

www.ingramcontent.com/pod-product-compliance
Lightning Source LLC
Chambersburg PA
CBHW070726220326
41598CB00024BA/3315

* 9 7 8 1 0 3 2 9 5 7 5 0 0 *